T0116185

The Underdogs

ALSO BY MELISSA FAY GREENE

No Biking in the House Without a Helmet

There Is No Me Without You

Last Man Out

The Temple Bombing

Praying for Sheetrock

The
Underdogs

CHILDREN, DOGS, AND THE POWER
OF UNCONDITIONAL LOVE

Melissa Fay Greene

ecco
An Imprint of HarperCollins*Publishers*

Page 337 is an extension of the copyright page.

The names and identifying characteristics of the Keith family of Iowa have been changed to protect their privacy.

Portions of this book previously appeared in the *New York Times Magazine, Reader's Digest,* and *Good Housekeeping.*

A hardcover edition of this book was published in 2016 by Ecco, an imprint of HarperCollins Publishers.

FIRST ECCO PAPERBACK EDITION PUBLISHED 2017.

Designed by Shannon Nicole Plunkett

Title page photograph © Nata Sdobnikova/Shutterstock, Inc.

Library of Congress Cataloging-in-Publication Data has been applied for.

ISBN 978-0-06-221852-0

17 18 19 20 21   OV/RRD   10 9 8 7 6 5 4 3 2 1

To a young man of true heart,
beauty, and grandeur,
Fisseha Solomon Samuel
'Sol'
Our son
1994 – 2014

CONTENTS

Introduction

I live with dogs: three small- to medium-sized ones relentlessly, and an extra one on weekdays. Everyone lives with dogs. Whether or not they belong to us; whether or not we mind being awakened by their twitching, snorting, and pursuit of dream-rabbits in our beds at night; and whether or not we are amused by their standing on the open door of the dishwasher after dinner in full view of our guests, licking the dinner plates clean, we are surrounded by dogs. Humankind has lived with canine-kind for a long time. By 33,000 years ago in Siberia, there may have been "domesticated dog food," according to archaeologists. By 10,000 years ago, "prehistoric dog-lovers" were burying their dogs in carefully prepared graves: one dog had been arranged to look like he was sleeping, one wore a collar decorated with deer teeth, and one had been put to rest alongside small-scale implements that might have been his toys.

Today many humans feel a magnetic attraction to dogs. A baby's attention is *riveted* the first time she sees a dog. Her breathing changes and her heart rate speeds up and the expression on her face seems to express, with wonder and delight: *Well hello! Do you live here, too?*

At the far end of a human life span, a dog is among the last things recognizable to a patient wasted by neurodegenerative disease. Memory for facts, faces, and autobiographical details may have been lost; intelligence, sense of humor, and creativity—gone; and the ability to stand, speak, sing, clap, or laugh extinguished. And yet, if a therapy dog is led into a nursing home to the side of a patient with dementia, it sometimes kindles a powerful reaction, as if the ravaged interior canyons held a few last exhalations of joy. *I'm so sorry but I just don't recognize you people standing at my bedside, showing me knickknacks, and weeping. But—oh my goodness!—who is this now?! Here is someone special! I know you.*

I've seen it myself, in a severely compromised child, during one of many visits over the last few years to 4 Paws for Ability, a nonprofit service dog academy in Xenia, Ohio, and a world leader in training assistance dogs for children with autism, seizure disorder, Down syndrome, attachment disorder, fetal alcohol syndrome, post-traumatic stress disorder, and other physical and "invisible" disabilities. To date, 4 Paws has placed nearly a thousand dogs in the United States and abroad. In class one month, there was a multiply handicapped boy of indeterminate age with severe cognitive *and* physical impairments. He was tilted far back in a massive motorized wheelchair and spent part of each hour on a ventilator. His mother, grandparents, and sister attended the class, as well as a full-time nurse who often took the boy's pulse and temperature and sometimes pounded on his chest, alarming us all. What could a dog mean to this child who seemed to tremble on a narrow rim of life?

A gorgeous black Labrador retriever was led into the room and across the training circle by Jessa Kenworthy, a young trainer (now a senior trainer) crisply dressed in khaki slacks, white sneakers, and a white polo shirt with the 4 Paws logo. The little boy's wheelchair had been sprinkled with dog treats. The Lab put two broad paws up on the arm of the wheelchair, leaned in to vacuum up the treats, and then snuffled in closer, moving up and down the boy's chest, neck, and chin with shallow sniffs. With a great bovine nose, she suctioned the almost-translucent little hands lying on top of the blanket. Under the oxygen mask, a pale ageless face contorted, the lips stretching sideways. We couldn't see the child's eyes. Was he expressing fear? Was he having a seizure? I looked from the mother on one side of the wheelchair, to the nurse on the other side, to Jessa standing back while loosely holding the lead and I wondered if someone ought to restrain the dog, protect the child. A weird whining sound came out of the boy's nose. But it wasn't a cry of terror. The child seemed to be wheezing . . . with laughter. The expression on his face—we could make it out now—was an enormous lopsided grin. He kicked his feet in mirth. In the circle of folding chairs, the rest of us laughed with relief and pleasure. The gleaming dog looked around at us, opened her mouth, and whipped her tail back and forth as if she were laughing, too.

Our bond with dogs is an ancient facet of our humanity: among the first passionate lifetime interests to be revealed, among the few identifiable features still intact where there is cognitive disarray, and among the last of our personality traits to deteriorate as we near the end.

SPELUNKERS EXPLORING A CAVE in the South of France in 1994 stumbled into a cavern of rock-wall paintings. The leaping, vibrant images would be dated to before 30,000 BCE—the oldest paintings in Europe—and would be hailed as among the most beautiful works of art of all time. La Grotte Chauvet-Pont d'Arc, the Chauvet Cave, is basically the Sistine Chapel of the Upper Paleolithic.

Nearly overlooked, fossilized on the floor, were markings less colorful but equally spectacular: 26,000-year-old footprints made by an eight- to ten-year-old child of the *Homo sapiens*. To this day, they are the oldest known human footprints in Europe.

The third remarkable discovery in the Chauvet Cave was this: alongside the child's prints appear the plodding footprints of a large canid. A Paleolithic wolf, a proto-dog, a wolfdog, a dog—whatever we call it now, the animal lived during the mysterious span of time when the evolutionary line of "dog" began to diverge from a now-extinct subspecies of *Canis lupus*, the gray wolf.

The path was narrow and the footprints may not overlap, suggesting that Kid & Canid strolled side-by-side, rather than leaving imprints in the mud at different times. The Wolfdog seems to have accompanied rather than chased the child, and at a calm pace, too, *not* like predator and prey, according to some expert trackers. And, given the absence of bones in the cave, the toothy carnivore seems not to have eaten the vulnerable little human. Of course it's conjecture, but one can imagine the pair of them: the child lifting a torch with one hand to take in the vibrant galleries while resting the other hand lightly on the bristly fur of the animal's back—for balance, or in friendship, or for reassurance in the pitch-dark recesses of the grotto.

EVEN THOUGH IT'S UNCLEAR *WHERE* it happened—East Asia? Europe? Russia? the Middle East? central Asia? Siberia? southern China?—and no one knows for sure *when* these events began—12,000 years ago? 20,000 years ago? 30,000 years ago?—and it's unclear why or how it started; and it's possible that the transition happened more than once, in more than one part of the world; and it's pretty clear that even as it was under way, there was backsliding and cohabitation with wild friends and relations, the domestication of dogs was the first event of its kind in the march of human civilization.

Often it seems incredible that the wolf-to-dog alchemy worked. We marvel that these keen-nosed sharp-toothed predators swapped out their pursuit of large game and their howling convocations under the stars for plush squeaky squirrels, monogrammed bowls, and a romp in the dog park. We do know that dogs were domesticated 20,000 years before sheep. Perhaps in another 20,000 years, we will have *sheep* standing on the open doors of our dishwashers.

An important current theory holds that dogs and humans

domesticated each other, that we *co*evolved, facing the precarious world of the Eurasian steppe (glaciers coming, glaciers going) first as parallel sets of hunters, then as fellow travelers, and then, somehow, as allies. "The emerging story," writes science author and environmentalist Mark Derr, "sees humans and proto-dogs evolving together: We chose them, to be sure, but they chose us, too . . . Both wolves and humans brought unique, complementary talents to a relationship that was based not on subservience and intimidation but on mutual respect."

"Man's best friend" doesn't really do it justice. A wild wolf subspecies was humanity's *first* friend, the first other animal to like us. Not many other species *do* like us, to this day, not of their own free will. (Cats . . . sometimes.) The wolfdogs of the Ice Age are like the preschool friends you still love decades later, the ones who attended your fourth birthday party and knew your grandparents, who remember when you wouldn't leave home without a pillow named Mrs. Squishy, and who always ran to join you on the playground at recess. (Cats are more like the middle school friends, who—if you cross paths with them in adulthood—may purr welcomingly and knock their heads against yours, or may suddenly grow alert to something in the distance, narrow the vertical slits of their eyes in concentration, and stalk wordlessly away.)

Wolves and wolfdogs and dogs knew humanity back in the days before we had institutions like marriage, religion, or personal hygiene. Wolves approached *us*—wolves reduced their "flight distance" from us—when *they* were the apex predators on the Mammoth Steppe of Eurasia, when *they* knew how to track and bring down the gigantic herding ungulates: the woolly mammoths and steppe mammoths and aurochs and woolly rhinoceroses whose painted images would charge across the cave walls. We rookies, newly arrived on the continent from Africa, hadn't a clue. We could paint the damn things, sure, but how to *catch* one? The wolves showed us.

We were in awe of the wolves. Ancient peoples like the Gauls and the Hittites would call them Warrior Gods, or Lords of the Animals. A Carpathian Mountain tribe would name itself the "Dacians," the Wolf-People. Wolves did not need *us*. But evidently some wolves began to find edible stuff near our encampments and to grow comfortable with ever-shorter distances between us. Maybe a few of them began to like us a tiny bit. We liked them, too, and we went nuts over their descendants.

BUT THEN HUMANITY GOT BUSY inventing agriculture, dynasties, literacy, and hairstyles, and—with exploding self-importance—forgot it hadn't always towered over all other creatures. Wolves remembered, but by then they were fleeing for their lives from an upgraded and armed Mankind 2.0. Hunted nearly to extinction, they would never recover a fraction of their ancestral lands from us.

For a terribly long time we denied our humble beginnings, first on the African savannah and then on the Eurasian steppe, the latter among the wolves and the wolfdogs. It got to the point where the only origin stories we humans told about ourselves occurred in celestial realms. Charles Darwin popped up in the nineteenth century to point out a few obvious family resemblances to primates, but we roared in protest and circulated caricatures of him with his sorrowful bearded head stuck on a monkey's body. The notions of prehistoric kinships, alliances, or friendships with animals seemed the content of children's stories. Like overnight millionaires, we pretended not to know anyone from the old neighborhood.

They say it's lonely at the top. And it *has* been lonely to consider ourselves the only thinking, feeling creatures on earth. We're a sorely isolated species, and our isolation increases meteorically as our industrial civilization drives untold millions of animals toward extinction every year, perhaps a third of all living species by mid-century. We're standing alone on the playground again, as when newly arrived on the Mammoth Steppe from Africa.

THE GOOD NEWS TODAY IS that a great many smart people in high places are reappraising, with respect and wonder, our ancestral, interdependent relationships with animals. The excavation of caves and burial sites and the sequencing of wolf and dog genomes are illuminating the primordial human/wolfdog bond in ways that make us feel richly entwined with earthly life.

Discoveries like the Ice Age footprints on the floor of the Chauvet Cave prompt us to recall our original standing in the natural world, when we viewed animals with fear, deference, admiration, gratitude, or friendliness. *We* were part of the food chain then, needing every friend we could get, millennia away from calling ourselves masters of the universe.

HUMAN CHILDREN ACT AS IF they never forgot any of it! Not our hardscrabble origins among the other animals, not how we used to look at and feel about the other animals. Children are *born* eager to make the acquaintance of animals, and as equals, too, not as the owners and the owned, the civilizer and the eradicated, the eater and the eaten, or the trophy hunter and the dead lion, elephant, or rhinoceros. "Now eat your octopus," says an offscreen mother to a dark-eyed three-year-old boy in a high chair at a kitchen table somewhere in Portugal. As captured on his mother's cell-phone camera, the winsome curly-haired fellow named Luiz tries to fathom what is on his plate. It's not a *real* octopus, right? It *is*? Where is his head—in the sea? Oh, it's at the fishmonger's? Did the fishmonger cut off all his legs, too, like this? But why? *Why?*

"So we can eat it, the same way we eat chickens," says his mother.

"Chickens?" cries the little boy in a high note of disbelief. "Nobody eats chickens! They're animals!" In a rising-and-falling singsong baby Portuguese, he sweetly lectures his mother: "*Octopus* are aneemals. *Fish* are aneemals. *Chickens* are aneemals. *Cows* are aneemals. *Pigs* are aneemals. When we eat aneemals . . . they die." (Still offscreen, the mother is now weeping.) "I don't like them to die. I like them to stay standing up." (By this point I'm choked up, too, and recalling a similar conversation with my own daughter years ago: me, the deceptive adult, pretending that we wait for chickens to die of old age before eating them; she, Molly, trying to wrap her mind around this bizarre approach to poultry: "You mean like someone just walks behind a chicken all day, waiting for her to drop dead?" Not buying the propaganda any more than did the little Portuguese boy, she stopped eating animals.) "Aneemals: we should take care of them and not eat them," Luiz says in conclusion. His doting mama agrees that they will never eat animals again and asks him to please focus now on his rice and potatoes.

"Seventy percent of children confide in their pets," Dr. Alan Beck, a pioneer in the study of dog behavior and director of the Center for the Human-Animal Bond at Purdue University's College of Veterinary Medicine, told me. "Have you ever seen the enthusiasm with which a five-year-old talks to a hermit crab?"

Before adults enter the scene with claptrap about the superiority of humankind, the grottos of children's brains probably look like the subterranean rock faces on which appeared the Ice Age *animals:* prey we chased, predators who chased *us,* distant species of the air, and (occasionally) wolf-forms, dog-forms. Young children are not that clear on the fact that they're no longer being stalked day and night by ravenous animals. When our son Seth was five, happily sleeping on the top bunk above his younger brother, Lee, I rearranged the boys' bedroom one day, thinking they would enjoy falling asleep side by side. My husband and I lowered Seth's top bunk to the floor beside Lee's for a happy bedtime surprise. Seth *was* surprised, but not happy. He was beside himself with panic and paced up and down the hall and could not sleep and fretted all night long and finally revealed the issue: with his bed now on the floor of his second-story bedroom in landlocked Atlanta, Georgia, he'd become a sitting target for sharks. There was no reasoning with the child. The next day we restored his bed's status as a top bunk, where he would sleep in safety and contentment for many years, without evident concern for the little brother snoozing thirty inches below, shark bait.

For children—not unlike the humans of the Upper Paleolithic era—it remains a revelation to behold new kinds of animals, an important precaution to gauge the intentions of carnivorous ones, a thrill to creep close to nonthreatening ones, and the highest-possible privilege and honor to be accompanied by a Wolfdog.

CHILDREN AND TEENS WITH PHYSICAL or cognitive impairments, birth defects, or behavior disorders can seem unreachable—unable to communicate, to learn, to master toilet training, to eat with silverware, to respond without unpredictable violence, or to make friends or show love. Sometimes, as a last resort, a Hail Mary pass, their weary parents bring home a 4 Paws dog.

The 4 Paws for Ability dogs are trained service animals, of course, but a friendship kindled between one of them (who has no clue he or she is a "service dog") and one of these children (who may not know that he or she is "disabled" or "differently-abled") can be as timeless, deep, and joyful as the bond between any other human/dog pair.

Despite everything that remains unknown about the mysterious 26,000-year-old path of footprints winding into the primeval art gallery of *La Grotte Chauvet-Pont-d'Arc,* let it represent, for now, the earliest recorded story of a Canid, and the Child who Loved It. Like this primeval duo, every youngster setting forth from Xenia, Ohio, today is newly accompanied by a dog who offers balance, friendship, and courage in the dark.

CHAPTER 1

Juke

In an Alaskan Eskimo village, a tall, weathered green house sways slightly beside the Bering Sea. From the rooftop, which is the last lookout point on this coast, the Erickson family whale-watches by day and stargazes at night. Sometimes the night sky silently explodes with the crimson flares and neon-green swirls of the aurora borealis, as if from a huge celebration just beyond the curve of the earth. Donna and Jeff Erickson, an Iñupiaq Eskimo mother and Norwegian-American father, have raised five sons here, whom they call "the Norskimos." The oldest three are married, with children and careers, and the fourth is in college. The youngest, Logan, is a teenager and with him came mayhem and grief. And a dog.

At age two, merry and sociable Logan Erickson fell under the shadow of autism, as if a beast loomed above him, cutting him off from his family. He lost his words and giggles, withdrew eye contact, and began flapping his hands and screeching. He has spent nearly a decade and a half in that state, practically mute, incompletely toilet trained, and prone to night wandering (with odd preoccupations in the night, like finding a fresh loaf of bread on a cooling rack in the kitchen, rubbing it into tiny crumbs, and very carefully spreading the crumbs in a fine layer across the living room furniture and rugs), and he has periodic explosions of rage or anguish so violent that he injures himself.

Logan is tall, gangly, and sweet, with shiny short-cropped hair, long, elegant hands, and a startled expression. He's usually quiet and compliant. But when a major fit is upon him, he runs into walls and bloodies his head to the point of concussion; he rips off his diaper and slaps the walls with feces. His father, a deep-sea fisherman, high school basketball and wrestling coach, town leader, and environmentalist, must drop whatever he's doing, wherever he is, and run to his son.

Only one other individual walks straight into the maelstrom of windmilling arms and screeches, toward the wrenched and wretched face, to protect Logan from himself: it's Juke, a yellow Labrador retriever trained by 4 Paws for Ability. He's a solid animal, with pale lion-yellow fur and pink-rimmed hazel eyes. His jowls draw his lips down into a dainty frown, while the lift of his fine eyebrows adds a touch of expectation to his face. When Logan begins to contort and thrash as if struggling to get out of an invisible straitjacket, Juke clambers on top of him, compelling the boy to lie down. Wearing an unflappable expression, Juke does not dismount (though he may be thrown off by the bucking teenager) and he doesn't abandon Logan, not even for food, water, or a bathroom break.

Juke was trained to respond to Logan's meltdowns. But no one trained him to hang on to the bucking boy hour after hour, ignoring his own needs. In awe of Juke's fidelity, Donna and Jeff bring him water and snacks; when they see the need to go to the bathroom rise to watery desperation in his eyes, they insist that he take a break. They tend to Juke, while Juke takes care of Logan. Sometimes a meltdown escalates beyond Juke's ability to manage it. When that happens, he runs to find Jeff, with the harried look of a man who has peeled away from a sickbed just long enough to call a doctor. "That's my signal to clock in," Jeff says.

Unlike the exhausted parents, Juke can predict when a huge

meltdown is approaching. If Jeff's sleeping, Juke pokes him in the arm with a wet nose. If he finds Donna first, he catches her gaze in a certain way—"he kind of scrunches his head down a little and whines, without losing eye contact," she says. "That's our signal." If the family is outside in the cluttered yard abutting the frosty shore of Norton Sound, the dog pokes his thick muzzle through the banister of the kitchen deck and stares down at them, relaying: *It's coming.*

JEFF ERICKSON IS A PROUD MAN with a high smooth forehead and a receding hairline, small intelligent eyes, and a brown goatee, who was born in Minnesota to descendants of Norwegian immigrants. His parents moved to Western Alaska as teachers when he was a child and his immediate family includes an adopted Iñupiaq sister. He and Donna met when they were children.

Soft-spoken, erect, and graceful, Donna displays the same gentle *noblesse oblige* in her warm, hospitable, stacked-to-the-rafters-with-provisions kitchen as at the airport, where she is the village agent and station manager for Bering Air. Donna learned to speak Iñupiat from her grandmother. Following tradition, she stitches parkas, boots, mittens, and hats from the skins and furs of caribou, fox, wolf, rabbit, and marmot. She prepares traditional dishes like "muktuk," a frozen whale-skin treat, and "tundra pies" like Cloudberry Pie. Her modern variations include Caribou Lasagna, Halibut Soufflé, and Moose Roast with Wild Cranberries. Like most of Unalakleet's 688 residents, of whom three-quarters are Iñupiaq, the Ericksons practice subsistence living: ice-fishing, hunting, trapping, and gathering greens and berries.

She and Jeff had a great marriage, a happy family. They took Ryder, Kaare, Austin, Talon, and Logan hiking, camping, fishing, and trapping; they went onto the ice, into the tundra, and out to sea. It felt like they laughed all the time, like when the water bed they installed, with tremendous effort, froze underneath them the very first night. Then Logan's autism unraveled it all.

When they see the S.O.S. on Juke's face, Donna and Jeff leap into action. If Jeff's at sea, Donna phones him and says, "Juke alerted," and Jeff will wheel around and steer for shore, counting on Juke to contain Logan as long as he can. Logan may be calm when his mother or father arrives; he may be hunched over on the floor twiddling his fingers and bobbing his head in the huddle of bony knees and elbows that is his twilight life. That just means the storm hasn't arrived yet.

Donna and Jeff know that Juke's alerts are not neurological readings. They don't mistake him for an MRI scanner or a doctor. He's a *dog*. He is a dog who, by nature, by training, and perhaps by love, is extraordinarily attuned to a boy.

A BIG DOG KNOWS HIS BOY is in trouble and races to gets help. Haven't we heard *that* story before?

Lassie. Lassie did that! Every day, every episode, Lassie ran to get help for Timmy and his "catastrophe-prone family."

But Lassie was a fictional dog! British author Eric Knight's 1940 novel, *Lassie Come-Home,* inspired seven postwar American MGM movies and the Emmy Award–winning TV show *Lassie:* broadcast from 1954 to 1973, it won its time slot every Sunday night at 7:00 p.m. for seventeen straight years and would become the fourth-longest-running prime-time TV series of all time.

After a melancholy, whistling theme song that hinted at something beyond our youthful comprehension—perhaps the brevity of childhood, or of a dog's life—the sable-and-white rough collie detected a danger to Timmy and intervened. She pulled the boy out of an abandoned house, a mine, a wildfire, and quicksand;

defended him against a bear, a bull, a rabid dog, an ostrich, a tiger, an escaped circus elephant, and a pair of bad guys planning to rob the local dairy; and tracked him when he wandered into the badlands, got his foot stuck between two railroad cars, failed to exit a badger hole, and floated away in a hot air balloon.

Lassie was America's furry heroine. A credulous public, especially children, believed in her. Wasn't there something in the way she loved Timmy that resembled the way Frisky, Brandi, and Sparky loved the boys and girls at home? Stretched out on the floor in front of our black-and-white television sets, we extended our arms over the backs of our napping dachshunds, poodles, and beagles, and felt certain they'd rescue *us*, if only our suburban backyards included abandoned mine shafts, runaway locomotives, and escaped circus elephants.

Experts rolled their eyes. "Needless to say, no dog is like Lassie," wrote Dr. Beck. "Even Lassie is not like Lassie."

Lassie helped move American dogs out of the backyard and into the house, according to animal behaviorist and author Dr. Patricia McConnell, but the show "set a ludicrously high bar. There was never an episode about Lassie peeing on someone's pillow," she says. "People [came to] expect that they, too, would have a dog they never had to train to do anything and who always did exactly what they wanted."

Jokes and one-liners at Lassie's expense have enjoyed an even longer run than the TV series, practically a subcategory of American humor. In the first panel of a famous *New Yorker* cartoon by Danny Shanahan, a drowning boy cries, "LASSIE GET HELP!" to the dog on the shore; in the second panel, Lassie reclines on a psychiatrist's couch, evidently recounting her issues, while a balding doctor takes notes. Dr. Beck's favorite cartoon shows a man saying, "Lassie! Bring me the Phillips-head screwdriver!"

Following the lead of experts who deemed Lassie preposterous, the generation of Lassie's children learned to make fun of her. Among the most mimicked of all television dialogues is this:

Lassie: *bark bark bark!*

Mr. Martin: "Lassie, what is it, girl?"

Lassie: *bark bark bark!*

Mr. Martin: "Oh no! Timmy's in the well!"

In fact, that dialogue never occurred because, in 591 episodes, Timmy never actually fell into a well.

Lassie humor ridicules the notion that an ardent, brilliant, devoted dog would really rescue a child day after day, week after week, year after year.

JEFF ERICKSON, HAVING TAKEN A CALL from Donna that Juke alerted, docks his boat, sheds his boots and rubber overalls, and huffs up the stairs to relieve the dog from his post. If Logan is already twisting and banging, Jeff straddles and pins him while Juke disengages and slips quietly aside. The dog could take a break now: he could go outside and run along the path he has carved in the snow visiting the other dogs in the neighborhood. But he stays nearby, keeping watch. Often, as Jeff tries to disentangle and calm the fraught boy, he and Juke make eye contact.

Later, when Logan has surrendered to sleep, Juke, Donna, and Jeff reconnoiter in the living room. The dog rolls back and forth on his back, blows out a few big snorts, and tries to balance on his spine for a moment, forming first the letter C and then a backward letter C. Donna and Jeff laugh, rejoicing with him that the crisis has passed.

Juke is their best friend. When he arrived from Ohio a few years ago, he stepped into a world of shocking loneliness and isolation, subzero temperatures outside the house and despair within.

He didn't promptly make the household shipshape. He was no furry Mary Poppins, primly certain that he knew best, clapping his paws to spur obedience. It took him a few months to get the picture. Little by little, he combined his innate dog behaviors with his learned service dog skills and modified both to meet his family's needs. The result is one of the "miracles" of the 4 Paws service dogs (though dog lovers all over the world attest to all sorts of miracles performed by their own dogs, professionally trained or not). Juke, like most dogs, seems to engage with his humans according to the promptings of his own mind and heart.

DO DOGS ACTUALLY *HAVE* "MINDS" AND "HEARTS"? Do animals think and feel? Remarkably, this has been a fertile subject of debate for hundreds of years.

At the dawn of our species, we assumed they *did* think and feel, that the nonhuman eyes peering at us from beyond the firelight, from across the savannah, or from up in the tree canopy glittered with intelligence and sensibility. An animal might appear fearful or curious, crafty or predatory, solitary or sociable, but each one seemed *aware*, possessed of free will, and not inferior to us.

Our Ice Age ancestors adorned rock walls with paintings of animals; our slightly less-distant forebears wove stories about an Animal Kingdom teeming with rulers and rivals, lovers and heroes, tricksters and ingénues. The respect of indigenous peoples for the wisdom of animals—including the quiet knowingness of reptiles, fish, and insects—embellished all of ancient art, literature, and music. Tribal legends bragged of the special associations of clans with specific animals, with the Hawk, or the Heron, or the Elk, or the Wolf. Human beings covered themselves with glory by advertising their special animal relationships, their Brother Bear, their Father Eagle, their Grandmother Tortoise.

BEGINNING WITH THE ANCIENT GREEKS, our admiration for the minds and hearts of animals faltered. Aristotle's fourth-century BCE *Scala Naturae* ("Natural Ladder") ranked all organisms according to "perfection." Man took the peak. Sponges and jellyfish clung to the sides of the base like barnacles. Barnacles also clung to the sides of the base like barnacles. The *Scala Naturae,* the "greatest biological synthesis of the time," remained the ultimate authority on the natural world until the rise of modern knowledge in the sixteenth century.

In the Middle Ages, the "Great Chain of Being" replaced the Natural Ladder as a diagram of the universe. Man still ruled over all other living things, but now God and the angels stood higher. With humanity as the crown of creation, nonhuman animals lost the esteem they'd enjoyed in earlier cultures. Now animals were judged to be incapable of thought or feeling, friendship or love.

Modern European philosophers like René Descartes in seventeenth-century France agreed, describing animals as "complex organic machines," without heart or mind, consciousness or volition. Animals "eat without pleasure, cry without pain, grow without knowing it," opined his follower theologian Nicolas de Malebranche.

"We place ourselves at the top of the evolutionary ladder," notes the eloquent modern chronicler of animal life Elizabeth Marshall Thomas. "Of course we do. We invented the ladder."

This disparagement of animals has persisted well into our own day.

IN THE TWENTIETH CENTURY, scientists began scanning the heavens for intelligent beings in outer space, as if looking for *someone* worthy of our friendship. A few other scientists, with more modest ambitions, began looking for evidence of intelligence in animals. From the start, the benefit of the doubt and the bulk of the research money went to the aliens, as if we were far more likely to find thoughtful individuals 50 quadrillion light-years away rather than nesting in Earth's trees, singing in Earth's rain forests, grazing in Earth's meadows, and sleeping at the foot of Earth's beds. Often under Earth's covers.

Edward L. Thorndike was a pioneer in the search for animal intelligence. A professor of psychology at Teachers College of Columbia University, he built "puzzle boxes" for cats out of scavenged wood and chicken wire. They looked like Cub Scout projects in urgent need of parental intervention. Then he gathered a few cats to see if any possessed a shred of "intelligence," a trait that has been broadly defined as "catching on," "making sense of things," "figuring out what to do."

Inside a puzzle box, a cat curled and shimmied about, meowing, flipping her tail, not thrilled with the confinement, and curious about the scrap of fish that had been placed just outside the box. The challenge was whether she would discover that pressure on the foot pedal would cause the door to swing open. If the cat made no progress, Dr. Thorndike gently placed her front paw on the foot pedal to see if that would inspire her. It would not. Cats prefer that strangers refrain from touching their paws. Sometimes he encouraged a cat to observe a second cat who'd mastered the foot pedal. That didn't help either, besides which it's difficult to encourage a cat to observe anything the cat isn't already looking at. But if the cat—while circling about and whining—accidentally stepped on the foot pedal, the action freed her! She gobbled up her scrap of fish. When Thorndike reintroduced her into the box, she escaped more quickly the second time and quicker still the third. This was a discovery of "operant conditioning," of the tendency of living creatures to repeat an action that leads to a pleasant outcome and to diminish the performance of an action leading to an unpleasant outcome. Thorndike called it the Law of Effect and it became a foundation stone of America's twentieth-century Behaviorist movement.

Had Thorndike found an intelligent cat? He believed *not*. Thorndike, along with his rough contemporaries Ivan Pavlov and B. F. Skinner, would become a founder of the Behaviorist approach to psychology, which was interested in what could be *observed*—i.e., behavior—and pointedly *disinterested* in invisible mental processes that might not even exist. Why go to such lengths as to postulate consciousness in animals, they asked, when a far simpler stimulus-response model could account for an organism's activity? For Behaviorists, animals were automatons who reacted to stimuli robotically, without thought, awareness, or problem-solving skills.

Not only did Behaviorism rule out animal *intelligence,* or "mind," but it also made the notion of animal *emotion*—"heart"—laughable. Emotions were the very essence of being human and *not* the essence of being orangutan, tiger, octopus, penguin, or dog.

Behaviorism would become the dominant psychological paradigm of the first half of the twentieth century, and its perspective toward animals would remain formidable through the end of the century and beyond. Mid-twentieth-century thinkers coined a nickname for the sentimental notion that animals could think and feel: they called it "the Lassie Myth."

"SCIENTISTS HAVE BEEN LED TO BELIEVE that only human beings have thoughts or emotions," Ms. Thomas lamented in 2000. "Of course, nothing could be further from the truth."

Jane Goodall, the legendary discoverer of the hidden world of Tanzania's wild chimpanzees, writes of being upbraided at Cambridge University for her lack of scientific method: "for naming the chimpanzees rather than assigning each a number, for 'giving' them personalities, and for maintaining they had minds and emotions. For these, I was told sternly, were attributes reserved for the human animal. I was even reprimanded for referring to a male chimpanzee as 'he' and a female 'she.' Didn't I know that 'it' was the correct way to refer to an animal?"

"Implying that similarities exist in the emotions or motivations of animals and people," wrote Dr. Beck in 1996, "is still one of the most egregious sins that a scientist can commit."

The overview of animals as unthinking automatons "has enjoyed curious staying power," Alex Halberstadt recently reported in *The New York Times Magazine:* "The notion that animals think and feel may be rampant among pet owners, but it makes all kinds of scientific types uncomfortable . . . 'If you ask my colleagues whether animals have emotions and thoughts,' says Philip Low, a prominent computational neuroscientist, 'many will drop their voices to a whisper or simply change the subject. They don't want to touch it.' "

UNDETERRED, IN THE PRIVACY OF their own homes, millions of dog and cat lovers, equestrians, birders, beekeepers, hikers, conservationists, nature photographers, aquarium and terrarium hobbyists, vegetarians and vegans, wildlife rehabilitators, and fans of nature documentaries narrated by Morgan Freeman and David Attenborough never questioned the sensitivity and intelligence of animals. Either they were unaware of science's disdain for the Lassie Myth or they didn't give a hoot.

Today, photos and videos of dogs, cats, and other animals displaying their feelings float through cyberspace. Humans click on, weep over, and share them by the millions. A sweet image of a baby elephant made the rounds, in which a petulant little fellow has pitched himself face-first into the mud while his concerned mother stands above him, watching. A caption written by a Kenyan park ranger confirms one's initial impression: "Elephant has subtle and rich feelings . . . Baby elephant will get upset when felt something wrong, just like human baby. Then it may express itself by crying or throwing itself into mud."

Recently I clicked on—and was captivated by—a brief video of a cat petting a pig to sleep. The cat, with sleepy half-closed eyes, purrs as he gently strokes his bristly-nosed friend, and the pig's fat nostrils twitch as he snores in contentment. It's had 10 million views. By the time I've finished replaying it, it will have had 11 million. The news outlet offering the video explained why they shared it: "Because sometimes you just need to see a video of a cat petting a pig to sleep."

More than *50 million* people have viewed forty-five-year-old footage of the reunion of Christian the lion and the two slim, long-haired, bell-bottom-wearing Australians who'd rescued him from a London department store as a cub, raised him to adulthood in a London garden, and then flew him to George Adamson in Kenya for supervised introduction into the wild. A year later, despite warnings that Christian had become a wild lion—leader of a pride, traveling with a mate—the young men flew back to Kenya for a glimpse of him. In the backcountry with Adamson, they spotted their favorite lion on a rocky hilltop but felt unsure that Christian would recognize them or want to be bothered. The golden beast prowled slinkily down the hill and then broke into a run straight into their arms. He stood on two legs, wrapped his front legs around the young men's shoulders, hugged and head-butted and nuzzled them, first one and then the other, again and again. On YouTube,

Whitney Houston's "I Will Always Love You" accompanies the almost-unbelievable reunion. Even without the song, the message couldn't be clearer.

As regards the inner lives of animals—and their capacity for friendship, love, and loyalty—popular culture got there well ahead of science. On the question of whether animals think and feel—unlike the subjects of evolution, or climate change—most "deniers" have been found in academia, not on the streets.

ONE DAY IN ALASKA, in Logan's special ed classroom at the village public school, Juke left Logan's side—which was unusual—and sought out another child. Logan's teachers have learned to pay close attention to the big yellow dog, so they watched when Juke crossed the room to a little girl sitting alone on the floor. He slid down beside her and put his head in her lap. She bent low over Juke, her long raven hair curtaining them off from the room, and began stroking and murmuring to the dog. After a while, Juke got up and returned to Logan. The next day Juke sought out the little girl again; again she curled over him and petted him; and the day after that he lingered even longer with her, whining a little.

While Juke has been trained in specific behavior interruptions with Logan, everyone acquainted with the dog knows that he is constantly sharpening his intuitions and expanding his range of interventions, so the teachers watched attentively. They didn't ask, *What's gotten into Juke?* They asked, *Is something going on with the other child?* It's not unusual for 4 Paws service dogs to alert to children other than their own. The teachers paid extra attention to the little girl and finally made inquiries. Within a day or two, they learned that there was upheaval in the child's home, a marriage breaking up, extreme emotions being expressed. A social worker was dispatched. While such details are far beyond the ken of a dog, Juke had evidently tuned into some wavelength of distress or sadness emanating from the little girl.

Is this a Lassie Myth story? Kind of. But increasing numbers of people around the world, including scientists, are finding it hard to deny that dogs and other animals display many of the hallmarks of thought and emotion. A growing body of evidence suggests that our profound sense of intimacy with dogs—our loving intuitions about them, their uncanny intuitions about us— is a result of our tens of thousands of years of shared life. Experts find less to laugh at in Lassie with every passing year. Where service dogs stand watch over fragile children, their parents don't ridicule Lassie at all.

CHAPTER 2

4 Paws for Ability

Odd break-ins were happening in a rural Cobb County, Georgia, neighborhood, twenty-five miles northwest of Atlanta. A narrow asphalt road separated two facing rows of forty-year-old split-level houses, many with chickens or goats in their backyards. People stood in small bunches on the road and talked about the break-ins. Nothing had been stolen yet, as far as anyone could tell, and "break-in" wasn't literally true, since nothing was broken in the entering, but there were clues: a basement door or a breaker box stood open. The back had been lifted off a toilet and unevenly replaced. A partial wet footprint shone briefly on a front hall. Then there was a ghostly sighting: a slim white leg and foot slipped out a quickly closing back door as the owners were coming in the front.

"What the hell?!" people said. This was gun-toting country. A few men made chuckling reference to what they might or might not do if they caught the intruder.

Halfway down the street stood the house of Mike Schwenker and Jennifer Jenka Schwenker: in her mid-forties, Jennifer is a round-faced, apple-cheeked elementary school teacher in granny glasses and a flowery top who gathers her wispy hair into a soft pouf; a sweet, merry, and lovably rattled type. Mike, in his early fifties, is an army and air force vet and aviation engineer with neatly clipped straight brown hair and a graying goatee, a fact-based individual who considers the data, then weighs in toward optimism if possible. In their house, the opposite problem was discussed. Nobody was breaking in. Somebody was breaking out.

Patiently, gazing aslant and down through rectangular metal-rimmed bifocals, Mike Schwenker installed window-pin locks, loop locks, security bar locks, and two-cylinder key-controlled dead locks at every window and door. He had degrees in aviation technologies and was a maintenance coordinator for Delta's Boeing 757 fleet, working beside air traffic controllers and dispatchers. Now he bent his mechanical know-how to preventing escapes. Having secured the house to his satisfaction—"It's like a prison," he told his wife—Mike latched his toolbox, stepped on a stool, and slid it to the back of the cabinet above the refrigerator.

He pocketed the door keys and gave a set to Jennifer, which she also pocketed.

But the same ethereal little presence spooking the neighbors darted soundlessly behind their backs, located the toolbox, delicately extracted the correct jeweler's-precision Phillips-head screwdriver or 8/9 open-end wrench, pussyfooted up the stairs, perched on a windowsill, dismantled a window block, slid open the glass, and, like Peter Pan, flew away.

Also like Peter Pan, he left behind his shadow.

The neighborhood interloper, Ben, had an identical twin brother, Sam, who had the identical diagnosis. *"Both?"* Jennifer had asked the developmental pediatrician when, in 2006, the three-year-olds were diagnosed with autism spectrum disorder (ASD). But, honestly, she already knew. Among her teaching credentials was special education certification. Her darling boys were textbook cases. They hadn't reached for objects, rolled over, or sat up at the expected times. They avoided eye contact. In church day care, a volunteer asked her if the babies were deaf. As toddlers, they overturned their toy cars and riding toys and spun the tires endlessly, hour after hour, both zoning out, hypnotized. When excited, they jogged in place, flapping their hands above their shoulders like hummingbirds.

Mike did *not* already know. He'd barely heard of autism. He thought it meant "mental retardation," severe cognitive disability. Did it mean they were becoming the people about whom others joshed: *The elevator doesn't reach the top floor? The lights are on but nobody's home?*

But once they started walking, the little boys followed their father on silent feet as he stripped floors and spackled walls and replaced wiring in their fixer-upper of a house. There were lulls in their self-stimulating behaviors when Mike took out his tools. They tracked the screwdriver or drill in his hand—from toolbox to task and back. The tools acted like magnets on the boys' eyes; the eyes silently slid, like nickels along a slot, following the arc of the tools' movements. The glint and precision of the pliers, ratchets, and nut drivers calmed them; the neat black indentations in the toolbox shelves, where each intricate steel or chromium thing fitted into its precise slot, pleased them. And when Sam Schwenker applied his mechanical intuition to filming his favorite thing—sock puppets—on his parents' cell phones ("You don't even miss your phone," Mike told me, "but when you pick it up off the kitchen counter, you see its memory is full—full of sock puppets"), and Ben Schwenker applied his mechanical sixth sense to undoing his father's barriers and locks, Mike knew they took after him. The lights were on and somebody was home.

Dr. Leslie Rubin, director of developmental pediatrics at the Morehouse School of Medicine and former director of developmental pediatrics at Emory University School of Medicine, diagnosed the boys. He encouraged Mike and Jennifer to love the children, to treat them as normally as possible, and to stimulate them with every possible experience. "There is no telling what they might achieve," he said. And meant it. "I don't diagnose children as having 'mild,' 'medium,' or 'severe' autism, unless the parents really press me on it, and then I err in the direction of less impacted," he told me. "Years ago, a mentor of mine said that we were not allowed to deprive anyone of hope, and that resonated deeply with me. If parents ask me, 'Will he go to college?' I say, 'I don't know, but let's shoot for it.'"

But keeping the boys enriched, stimulated, and *safe* was a tall order. Ben, like an estimated half of young children on the autism spectrum, is a "runner" or an "eloper."

Where is Ben *right now*? his parents asked themselves with every breath. If one checked an iPhone, the other watched Ben. If one stood at the stove to cook, the other watched Ben. When Mike went out to work, Jennifer watched Ben all day. But, once in a great while, when they both looked down simultaneously at their phones, or let their thoughts drift outside the locked-up house, or Jennifer sagged into the sofa on an interminable afternoon and closed her eyes for a moment, Ben made his move. "You can be a few feet from him and he gets away," Jennifer told me. "He watches your gestures and intentions. He studies the whole situation, like a squirrel figuring out how to get in your bird-feeder."

In the blink of an eye, the scrawny kid perched on an upstairs windowsill, screwdriver in hand, liberating himself. He dropped to the ground—his thin flop of straight brown hair ballooning briefly in the air—and took off across the backyard, pulling off his clothes as he ran. He scaled the tall wooden backyard fence that had been built to corral him and his brother, the fence he'd mastered by age four. Naked, he disappeared into a no-man's-land of fields, subdivisions, strip malls, and state routes.

Typical of children with autism, Ben ran first toward the glitter of water. He stripped as he ran and leapfrogged fences because, regardless of the season, he planned to go swimming. He and his brother had been taught how to swim; still, these disappearing acts put Ben at great risk. Not only would he be swimming alone, but he was clueless about guard dogs, strangers, traffic. Did he even know his last name and address, in case someone wanted to guide him home? If he knew those words, he'd never said them aloud.

He skinny-dipped as the uninvited pool guest of some unknown neighbor, climbed out, and found his way into the house. He wanted to examine wiring, HVAC units, laundry vents, attic fans, and pipes. Residential, industrial, and automotive systems called to him. "He wants to see plumbing, and floor plans," says Mike.

"It's like a jones. His heart goes a mile a minute. When he gets to look at new stuff, he gets an endorphin rush."

Neighbors were mystified. Then a local couple nearly tripped over the culprit: they came in one night to find a skinny, wet, naked little white boy sprawled out on their kitchen floor with his head inside the cabinet under the sink. He appeared to be studying the garbage disposal unit. They yelped and demanded an explanation, but the boy ran away. They pursued him out the door and soon came across the Schwenkers, jogging down the road and calling, "BEN!"

Far worse for the Schwenkers were the times that *no* neighbor emerged with a complaint. Then Mike and Jennifer froze in the driveway, not knowing which way to run first, willing themselves *not* to hear a squeal of tires in the distance or the revving-up whoop of a siren. They tottered about—a few steps this way, a few steps that way—looking at each other with stricken faces. They dialed 911. They dashed back into the house to secure Sam and keep him with them. The universal parental monotone prayer-to-avert-disaster moved on their lips: *Please please please please* or *Oh God oh God oh God*.

Mike worked long hours at Delta, including nights, weekends, twelve-hour shifts. Other than the few hours per day the boys went to school (to special ed classrooms from which they came home frazzled and incoherent), Sam and Ben and Jennifer stayed sealed up in the silent house on the quiet street. Locked inside the house all day, the boys "stimmed" (self-stimulated through repetitive movements), kicked, pinched, licked, or slapped each other; they literally bounced off the walls. Jennifer fell into silence and claustrophobia; she was deeply alone. The neighbors she knew left for work in the mornings, so there was rarely even a car rolling down the road. Some days the mailman was the only adult she saw, through a locked window. Sometimes her optimistic voice fell flat. Words like "Let's build a blanket-fort—that'll be fun!" sounded false even to her. Though she resisted, she sometimes gave in to moments of weeping—alone in the bathroom, hiding her distress from the boys. There were days she hardly combed her hair or put on an attractive pair of slacks. Every expert agreed that the boys needed real-life adventures and experiences outside the house. But it mostly fell to Jennifer to take them.

As trapped as she felt in the house with Ben and Sam, she actually didn't feel *less* lonely in the car with the two of them yelping or peeing or punching each other in the backseat. Other drivers, sealed up cheerfully alone in their vehicles, might as well have been aliens floating beside her in soundproof space vehicles.

The boys' favorite outing was to Best Buy. Sam, ever on the lookout for puppets, pulled his mom toward the big-screen televisions in case *Sesame Street* was on, while Ben was desperate to reach the appliances. When Mike was available, they made good use of their time by splitting into teams of two. But when Jennifer, because they had to go *somewhere,* took the boys alone, she could end up shipwrecked: arms outstretched in the midst of the cell-phone department as Ben, with all his strength, tried to tow her toward Large Appliances, and Sam threw his weight into dragging her to TV & Home Theater. And it did no good then to ask herself, *What was I thinking?* because she knew she had thought: *If we don't leave the house today, I will lose my mind.* Of course by then they were becoming a public spectacle—both boys beginning to freak out, to squeal, to fall down and rage, squirming across the tired carpet, each in his longed-for direction—and it felt as if she'd never manage to get them home. "Can I help you?" an employee might kindly ask, or sometimes even a nice shopper who sensed her distress, but no one could really throw a rope to her, no one who didn't know her children, or didn't know autism, or couldn't begin to fathom why she would have entered this buzzing vibrating fluorescent-lit electronic playground in the first place. A stranger's awkward attempt to weigh in could catapult one or both boys into the outer rings of hysteria. She had to grapple her own way back to the car with the boys, and then spend most of the next hour trying to get them to simmer down and allow her to buckle their seat belts.

It was a rare excursion that ended peacefully. By the time she staggered, dragging them, back across a parking lot, her hair was undone, her face red, her dress askew, and the boys were wailing and unhinged, their brief joy turned to confusion.

Could anything be worse than these attempted field trips to appliance stores? Yes. There were parks and playgrounds and wooded paths all over Atlanta, but none with escape-proof perimeter fencing. This she learned one day as she watched Sam zip down a playground slide and waited a beat for Ben to appear right behind him. But Ben didn't come sliding down after his brother and Jennifer couldn't find Ben elsewhere on the playground equipment. Then she spotted him in the distance, hot-footing it toward the street, shedding his clothes as he ran.

"I'M ACTUALLY TURNING INTO A SHUT-IN," Jennifer told her husband as gently as possible, at a moment she thought he could bear to hear it. "I can't hardly leave the house. I can't hardly stay home anymore. It's like I keep thinking: *This is my life*." She followed up with a soft laugh.

They couldn't afford a nanny; she knew they couldn't. They couldn't afford a babysitter. They couldn't afford private school or private tutors. And there was no end in sight—not high school graduation, not college graduation, not young adulthood, not marriage, at least not at the rate they were going.

Mike did everything possible to make their house an oasis. He arranged for the delivery of the cockpit and fuselage of a discarded twin-engine Beechcraft Travel Air plane. He scrounged instrument panels, lights, switches, and seats from junked aircraft and he and the boys rebuilt it together as an airplane-playhouse.

Jennifer researched autism endlessly. She joined online associations of "autism parents." In early 2009, when the twins were six, she read about a child showing improved day-to-day functioning with an "autism assistance dog." She was intrigued and mentioned it to Mike. They loved dogs. Could this be an option for them?

He took it one step further: "I'll tell you what we need. We need a *tracking* dog. Is that a thing? Can you own your own tracking dog like the ones the police have for searches?"

"Oh my goodness, can you *imagine*?" sighed Jennifer.

EVERY MONTH, AFTER ROUGHLY a year of waiting and fund-raising, a dozen families pull up in rental cars from the Dayton or Cincinnati airport or in wheelchair-accessible vans filthy from cross-country trips. Xenia is a leafy antebellum village, a former safe haven on the Underground Railroad. A few blocks from downtown, on three acres of land, sits the former VFW hall, an aluminum-sided one-story rectangular building. It's a modern dog-training academy now. In the parking lot, the families unload their gear: little crutches or small motorized wheelchairs; walkers and gait trainers; aerosol equipment like breathing machines, mouthpieces, masks, and suction pumps; apnea monitors; floor mats and wedges; feeding tubes, feeding bags, and powdered formula; commode chairs, catheters, and diapers from small to adult sizes. There is a twelve-year-old boy with paraplegia as a result of spina bifida and another twelve-year-old boy with dwarfism, epilepsy, and developmental delays. There is a couple from a U.S. Army base in Michigan whose two young children were both just diagnosed with autism. There is a teenager with deafness. There is a teenage girl with brain damage and seizures as a result of violent early abuse by her birthfather; her adoptive family has brought her here for a dog. Many of the children engage in self-stimulatory behaviors like rocking, blinking, humming, hand-flapping, tapping the ears, rubbing something against the skin, licking or squeezing things.

The arriving families are white, African-American, Asian, and multiracial. Deeply weary—and not just from the journey from across the United States and from a few foreign countries— these are people whose lives revolve around keeping "special needs" children alive, and safe. In some families, a child's life-long crisis began with an accident or with the onset of a disease; for some, it was a genetic syndrome confirmed soon after birth; for others, there were red flags during pregnancy; and for others, a child by adoption began to reveal issues for which the parents felt unprepared. Some parents have not slept through the night since their child's first seizure, out of terror that the child will die alone in the night; they're sleep-deprived to the point of hallucination, even to the point of physical illness. Some families have not enjoyed a whole-family outing—forget the movies or the lake; they can't go to a restaurant—since the days their most-challenged child outgrew his stroller and became capable of staging traffic-stopping tantrums in public. Some of these children have never played in their own yards or hung out with other kids. Some of the mothers, pressed into lifelong service as therapeutic caregivers, are profoundly lonely and depressed.

Many families include siblings of the special needs child, often too young to understand why they *never* get to be the center of attention and why even the new *dog* will not be theirs.

Some of these families have been cheated by unscrupulous operators. Service dog training is an unregulated industry. "There's no national standard, no national certification, and no guide to reputable agencies," Karen Shirk, fifty-two, the founding director of 4 Paws, told me. "We end up picking up the pieces." Because parents of special needs children are often exhausted and economically strapped, they're vulnerable to hucksters advertising miracle-working dogs on the cheap. "Some agencies don't know what they're doing," said Karen, "and others are actively scamming people, like they place barely trained animals and call them 'companion dogs.' They promise things a dog could never do. One family was told their new dog would stop the child from eating dirt. *Really?* Dogs eat feces, bugs, vomit, and garbage, so how's a dog supposed to know which of thousands of food items a child shouldn't eat?"

Wheeling, carrying, coaxing, dragging, or chasing their children across the gravel parking lot toward the double glass doors, many parents experience a sudden flash of hope. No one is unreservedly optimistic. Everyone is well acquainted with cycles of optimism, disappointment, exhaustion, and despair. But as they spy handsome dogs frolicking on green acreage beyond the fence, some permit themselves the thought: *This is one crazy idea that just might work.*

THROUGH ONLINE FRIENDS, INCLUDING ONE Atlanta family, the Winokurs, the Schwenkers learned about the Ohio-based nonprofit, 4 Paws for Ability. The dogs were trained at a cost of $25,000 each, toward which a family was asked to contribute $13,000, with the difference made up by grants and charitable donations. The Schwenkers applied and were accepted; their church friends and Mike's airline colleagues pitched in, and they hit their fund-raising goal in four months.

Photos of Barkley, a Labrador retriever/bloodhound mix, were emailed to the family. He was a young dog with a hangdog face and pendant ears mounted on a sleek athletic body with a tall whip tail. The combination struck Jennifer like a lively small boy wearing horn-rim eyeglasses. Barkley looked earnest and well-meaning to her.

In September 2009, the Schwenkers drove to Xenia for the mandatory ten-day lcass at which they would meet Barkley and learn to work with him. In typical neighborhood dog-training classes, people learn basic commands from a trainer and teach them to their dogs. At 4 Paws for Ability, people are introduced to beautifully trained dogs. They must try to master a new language and begin to grasp what is possible, what they have been given.

NEWCOMERS PUSH THROUGH THE FRONT DOORS and walk smack into a *smell*. Everything reeks wildly of dog, with undertones of ammonia and a touch of the barnyard. Tufts of fur drift from room to room like milkweed seeds bobbing above a meadow. Though 4 Paws usually has 450 dogs in various stages of development, most live in foster homes around the region for maximum enrichment and socialization. Still, a few dozen live under this roof during periods of advanced training and family classes. But there's no forlorn racket of barking like the panicky bedlam inside a county animal control unit, because the dogs here enjoy busy lives and are not yelping for escape or rescue. The mood seems more like a kindergarten: some trainers work one-on-one with dogs, others lead small-group activities; half a dozen dogs are lining up to go out the back door for playtime or on field trips, and others are just coming back in, refreshed.

In a middle room, crates and pens temporarily hold young golden retrievers, black Labs, German shepherd dogs (GSDs), rough collies, goldendoodles, papillons, and more, including donated dogs, rescued dogs, expensive scions of British or northern European lines, and puppies produced by in-house breeding programs. Dog walkers, vets, groomers, local foster families, college students fostering puppies on their campuses, and visiting Scout troops come and go through a side door.

A cat lives here. She is a one-feline proof of the theorem that if a cat doesn't flee, dogs won't chase. She moves at her leisure along counters and the tops of crates above the dogs' heads. If she felt like it, she could walk among them—very slowly, very calmly, tail up, *not* fleeing, relaying with body language, *Let's not get excited, boys*—and arrive at her destination unmolested. A second cat appeared one day, looked around, and took up residence. Neither cat flees, so the dogs don't chase.

PRESIDING OVER ALL THE SOFT-FOOTED CHAOS is Karen Shirk who—like her clients—comes from deep in the trenches of disability, isolation, and depression. She was rescued by a dog. Nevertheless, she and her staff warn the newly-arriving families *against* the Lassie Myth. They know that, among the eager parents and veterans pushing into the social hall, there will be some who expect that—without much effort on their part—a brilliant dog will gallop into their lives and set everything straight.

To welcome families to the school she founded in 1998, Karen places a fingertip over the metal button of a tracheotomy tube in her throat, enabling speech, and speaks in a voice that is winningly husky. Positive changes *will* happen after placement, Karen tells a social hall full of nervous families, but the improvements will occur incrementally, and *only* if the adults in the household stay diligently engaged with their new dog, rounding out the training. When Karen talks, she lays a hand on her chest and reaches up with one finger to cover her trach. She happens to be a truth-teller, of the unvarnished sort, so the hand-over-heart posture is not only clinically required, but seems to underscore her delivery.

Most 4 Paws dogs are trained for children; about 5 percent go to military veterans with post-traumatic stress disorder. Every dog receives 500 hours of training, far above the industry average of 120 hours. "Mobility assistance dogs" are trained to open doors, turn on lights, bring the backpack. "Seizure assistance dogs" and "diabetic alert dogs" notify parents of the onset of a medical episode in a child. (Many predict incidents six to eighteen hours in advance, with about 80 percent accuracy.) "Autism assistance dogs" and "fetal alcohol spectrum disorder (FASD)/drug exposure dogs" learn a variety of "behavior disruption" commands to help a child regain self-control, including "Lap," for lying across a child and offering the comfort and stability of deep pressure; or "Disrupt," for gently nudging a child's hand away from self-harm like yanking out his hair or pounding his head with his fists. "Hearing ear dogs" respond to verbal commands and to hand signals; they make physical contact to alert their humans to doorbells, phones, and fire alarms. "Multipurpose assistance dogs" are trained to address a wide range of issues. For children who "elope," like Ben Schwenker, dogs are trained in tracking.

For children incapable of being the "handler" for their own dog (which includes most 4 Paws children), a system called "tethering" has been devised: from the dog's harness, a leash runs up to the parent, who is the true handler, while a second leash runs to the child. That second leash keeps a child—especially a runaway type—close to the parent, while giving the child a thrilling sense of freedom and importance. *I'm walking my dog!* The dogs are taught, by pairs of trainers, to remain calm and to focus on their adult handler while anchoring a restless little kid on the far end of the extra leash.

"In addition to doctors' and psychiatrists' and psychologists' reports, we need to see videos of your child," Karen tells all accepted applicants. "We want to see your child waking up, interacting with the family, eating breakfast, getting in the car, walking into school, interacting with other kids, making sounds, throwing fits. What sets off your child? What's the hardest part of the day? What do you need the most help with?"

Training Director Jeremy Dulebohn, in his early forties, studies the videos and medical files in order to set up kid-specific, family-specific dog training. One staffer will act out a child's behavior or medical issues, including seizures, tantrums, self-harm, or eloping. A second staffer teaches the chosen dog what to do in that event: run and get help, sit and bark, push close to the child to interrupt the behavior, or take off and find the child. The dog is rewarded for each step along a spectrum of increasingly appropriate responses.

Actually, the training starts in early puppyhood. There is an outbuilding made over into an indoor puppy playground, with stimulating activities for the younger set, including chutes and ladders, wading pools, and visiting animals and people. Identified in the 1950s by Dr. John Paul Scott, a researcher in genetics and animal development, there is a "critical period" in every dog's life—roughly the first fourteen weeks—during which the puppy is game to encounter new experiences and friends, and after which novelties are much scarier. By about fourteen weeks of age, a pup's responses to the world will be largely fixed.

The 4 Paws staff is on the lookout for what gives a puppy pleasure. "We ask, about each dog, what motivates him? What is he good at? Does food make him the happiest, or toys?" a senior trainer, Jennifer Varick Lutes, told me. "If the dog is great using his nose, then we think about steering him into tracking. If the dog is attuned to sound and notices small differences, she might make a good hearing ear dog. Really sharp with the nose and observant about personality, maybe a seizure assistance dog. We try to figure out every dog's personality and passion. We get clues from their foster homes, too: 'She always has her nose to the ground' or 'She loves to carry my keys around.' We want to find a task that the dog will love more than anything in the world, not just train for what the dog *has* to do. We try to figure out which dog possesses the innate behaviors that would be the most useful for a family. Rather than squelch a dog's nature, we look for the best possible role.

"We had a golden retriever who showed real promise in tracking. We started him in beginning nose training and he seemed to enjoy it. Jessa, another trainer, would hide, and we'd tell him: 'Find Jessa!' And he'd start out in the right direction but then he'd veer off to pick up a piece of trash from the ground. When he spotted a paper cup, his eyes would light up, he'd grab it, bring it back to us—he wanted us to have it. He was so proud of himself and so happy. We had to relay, 'No, we're not looking for a paper cup. We're looking for a lost person. Find Jessa.'

"He had such a good nose that we tried a few more times. I saw Jeremy coming back from field training with him one day, holding like four old paper cups and six empty soda cans. He said: 'This is not a tracking dog. This is a mobility dog.' So we trained him to do things like bring the newspaper and answer the phone. He loved it! He's a super mobility dog now.

"We also figure out each dog's learning style in order to shape the training method for him or her. You want to *create* a certain behavior and then *reward* it. You teach different components of a behavior set and then link them together. We note how a dog displays stress or fatigue and we relay that to the family, too, so they can keep an eye on how their dog is doing every day, whether the dog needs a break. None of this works if the dogs aren't happy, too.

"The whole thing at 4 Paws is that we see it as a partnership. It's not just what does the *client* need. It's about how a dog can meet those needs in ways that are really fun and rewarding for the *dog*. Dogs love having important work to do, and being needed and involved every minute of the day. They flourish in their families."

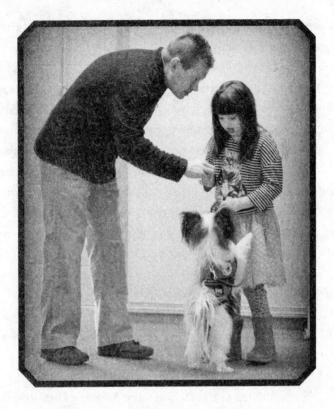

"THESE ARE *REAL* DOGS, we're giving you real dogs," Jeremy Dulebohn tells the new class. He stands in the center of a circle of sofas, old La-Z-Boys, and canvas sports chairs in the linoleum-floored old VFW social hall. Families have set up camp, arranging wheelchairs, floor mats, and pediatric medical equipment within the circle. A carpeted play area against a far wall offers a place where all the children, but especially those with behavioral issues, feel free to watch TV, play video games, do art projects, play with or break toys, pound on things, roll about, and shriek. At all hours, children and teenagers rock to and fro on two big plastic spring-suspension horses. The seesawing springs twang in the background all day.

Jeremy has a fit, self-contained, quiet bearing, and a kind of naturally skeptical watchfulness. While offering *his* pitch against the Lassie Myth, he offers a dubious half-smile and brushes his palm forward along the bristled top of his crew cut, as he tends to do when delivering a bit of regretful news. "It takes time, patience, and work to create an effective service dog," he tells the roomful of hopeful parents. "Lassie was played by a *dog actor*. There was a Hollywood trainer standing behind the camera telling Lassie when to sit, when to bark, and which way to run. Those weren't true stories. Lassie wasn't doing all that on her own."

The clients are nodding. Some take notes. They hope Jeremy is lying.

"I WAS *FLOORED* WHEN I met our first families," Jeremy told me. "I drove the hour and twenty minutes home in total silence. What these families deal with on a day-to-day basis . . . it's mind-boggling. I just had no idea. Kids screaming, throwing fits, wetting themselves, stimming, banging their heads, attacking their parents, or just zoning out; and kids where you never know when they're going to have the next seizure or when their blood sugar's going to crash or when they're going to stop breathing or when they're going to wander off and get lost; or kids with too many issues to be captured by one diagnosis. Looking back, I realize I wasn't always the nicest kid growing up. I teased other kids who weren't perfect in some way. Now I'm on the other side of it."

Jeremy is the matchmaker between dogs and families. He knows every dog in the pipeline and closely studies every family's application materials. "While I'm reading through a file and watching the video," Jeremy told me, "a dog will just pop into my mind: 'This dog kind of reminds me of this family' or 'This dog kind of resembles this family.' Karen and I usually start out with four to six dogs as possibilities for a family. As we get closer to class time, we make pencil marks through all except the dog we feel is the best match."

In Karen's eyes and in the view of most parents, he's a wizard. Still, roughly 10 percent of the 4 Paws placements fail. Dogs are not one-size-fits-all. They're not medical appliances that can be scaled up or down or refitted with a different tail or snout. No handle can be turned clockwise for greater output of energy, watchfulness, or barking, or counterclockwise for less. Dogs form their own opinions. Sometimes a dog bonds not with the child with special needs, but with the child's sibling or mom or dad.

"Some placements fail because parents aren't prepared for how much extra work a dog will be," Karen told me. "I understand: they can barely get themselves and their special needs child out the door in the morning; adding a dog—having to go through a few weeks or months of adjustment—feels overwhelming."

A placement may fail because a family didn't accurately relay the extremity of their child's behavior. The family may have hesitated to present the whole picture for fear of being turned down. "Some of the kids—when I see them in person—you know, they're not what we expected from the videos," Jeremy said.

Karen said, "Parents will tell us their son is 'gentle' and the child looks gentle in his videos, so we place a soft dog. But then the child's violent meltdowns scare the dog and the dog starts avoiding the child and they don't bond. None of this works if the child and the dog don't connect."

"One child kept pinching a dog's ears until they were swollen," Jeremy said. "It was like a nervous tic. We brought back the family and worked with the parents on extinguishing the behavior, but the child couldn't stop. We took back the dog." Karen said, "A year later, after the child's behavior was extinguished, we placed a new dog, the parents kept a close watch, and they did fine."

Sometimes 4 Paws realizes that no dog can be safely placed with a child or family and it acts to protect its animals, refunding the application fee and denying the application. "No child or family or other pet would be at risk *ever* from one of our dogs," said Jeremy, "but one of our dogs could be at risk in a family."

"Seriously, someone just asked me for a service dog to wake her up if she falls asleep while she's driving!" said Karen. "Oh my God! Okay, I understand narcolepsy is a bad problem to have, but it isn't safe for her or for anyone else, including the dog. Is the dog supposed to wake her up just as they crash into a tree? Had to say no to that one."

The 4 Paws staff does rigorous background checks, following up with every reference. If in doubt, a home visit may be paid. Karen said: "One reference I phoned for a family said, "Okay, well, don't quote me, but did she tell you her son kills frogs and small animals?"

JEREMY HAPPENS TO OMIT, DURING his anti-Lassie remarks, that (like Timmy) *he* has a canine sidekick whom he never leaves home without, manfully loves, and trusts with his life. Brody, a huge, shaggy black-and-tan German shepherd dog, naps in Jeremy's office just off the social hall. When Jeremy speaks his name, Brody shows up instantly, shakes himself off, and reports for duty. When Brody swaggers beside Jeremy down a sidewalk in town, people stumble trying to get out of the way.

Brody is subdued and gentle with children, offering an occasional quiet wag of agreement and a softening of the ears when little kids gather and ask permission to pet him; but, if an adult male were to threaten Jeremy, the dog's hackles would bristle and his head would drop and extend in a stalking stance. Jeremy wouldn't even need to speak. The dog wouldn't even need to bark, the low gurgle in his throat warning enough.

One day in class, someone mentioned the tragedy of millions of abandoned and euthanized dogs in America. Off-script, in a rare emotional moment, Jeremy suddenly said: "I mean, if I ran out of money, I'd lose my house before I'd give up this dog. I'd move into a cardboard box under a bridge before I'd give up this dog."

"Our goal is for the dog to bond with your child," Jeremy tells the families. "Your child should be the one feeding the dog treats. If your child is *not* able to give the dog treats, then it's your job to hide the treats on and around the child, on the wheelchair or in a pocket. You want the dog to feel: *I don't know what it is about this kid, but whenever this kid is around, good things happen.*

"We use positive reinforcement with the dogs," Jeremy explains to each class. "You'll hear people say you need a firm, authoritative tone with a dog, but it's not true. Imagine if I wanted my ten-year-old to pick up his toys and, out of the blue, I screamed at him: 'PICK UP YOUR DANG TOYS!!!!' It would scare the heck out of him—he wouldn't know which way to run—but it wouldn't *teach* him anything. But it also wouldn't work if I said: 'Umm . . . excuse me, but . . . do you think . . . maybe . . . if you have time later on today, would you mind very much picking up your toys?' That's just confusing. That sounds like something he can ignore. You want to strike a tone in the middle with your dogs. Don't use too many words. You don't need to say, 'Biscuit, will you come over here and let Mommy give you a big hug?' Just say: 'Come.' "

"Can she shake hands?" a new mother called out on the first morning of class one month, as Jeremy displayed a few maneuvers with one of the dogs.

Jeremy paused, tilted his head, and said, "Well, sure, she *can*, but that would be kind of like asking a graduate student to sing the alphabet for you."

"Something's wrong with this dog. He won't listen to us," a disappointed Alabama father of a special needs child complained one morning. The dog had spent the previous night with the family for the first time, at a Xenia motel. Now the father wanted his money back.

"Show me," said Jeremy, who knew the dog had been exquisitely trained.

"Dandy," said the mother in a Deep Southern drawl. "Now see-uht."

The dog stood waiting, wagging and expectant, for a command he could understand. "See-uht" didn't sound like anything he knew how to do. That prompted two changes in 4 Paws methodology: every dog is videotaped performing its entire repertoire in advance of a family's arrival—kind of like a manual accompanying the installation of new software. And every dog is trained to respond to commands in the way a client will speak them. Recently, for a teenager with deafness, a dog was taught to lie down on his blanket with the word "Pace," rather than the usual "Place," because that's how the young man pronounced the word.

"The dogs are nonjudgmental," Jeremy tells his classes. "You've got a kid who's picking his nose? The dog isn't thinking, *THAT is gross.* He's thinking, *Save one for me!* Or your child has disappeared and you say: 'Find Jeffrey.' The dog isn't thinking, *Jeffrey's in danger!* The dog thinks: *Game on!*"

FOR THE SCHWENKER TWINS, Jeremy had chosen the bloodhound/Lab mix and trained him in autism assistance, tracking, and tethering. If Ben got away, Barkley would pursue him. But as long as Ben was clipped to Barkley, his chances of escape would be greatly diminished.

On the family's first field trip in Xenia with Barkley, the great gleaming beast compliantly waited beside the 4 Paws van in a shopping mall parking lot while three leashes were attached to his harness: one sloped up to Jennifer and two hung sideways to the kids. Mike prepared to follow behind. Being outside on a windy day with his new family, being the center of everyone's busy attention and fresh rigging, seemed to please the big dog. Then the Schwenker family—like four planets in orbit around a shiny black sun—marched into the mall.

Tethered to Barkley, the boys tripped and tangled and double-stepped down the wide polished foyers. They seemed unaware of the fact that they were now—like water-skiers—being towed. If they tried to dash away, Barkley, like a puttering motorboat, churned forward and tugged them back into place.

After a few days of practice on the daily field trips, the children weren't just stumbling alongside their ox-like friend, they were putting out their hands, patting his broad back, and hanging on to his tail, almost as if they'd always known him. Ben, naturally, was interested in the mechanics of the thing. He liked to raise the flaps of Barkley's ears to peer inside, and when the great dog yawned, Ben positioned himself to get the long view straight down the throat. "We figured they weren't going to have the typical boy-and-his-dog kind of relationship with Barkley, but it was enough that they accepted him," Jennifer said. Their triplet dog brother.

Back in Atlanta, Jennifer prepared for outings like a para-chutist confirming every strap and rope before stepping into the air. It took a while to get the leashes and clips right; Ben mastered them instantly, escaped from the rigging, and ran away, but finally his mother outwitted him with a heavy-duty lock-ing carabiner. He might get away from her but he would *not* get away from Barkley. Jennifer felt happy. Even the time she spent getting Barkley and the boys ready to go outside was forward-moving time, the windy kind of time that connects you to the world rather than the old kind of time in her house that seemed to turn sepia from inactivity.

It was embarrassing, at first, to step into public. "We were a total spectacle, like a traveling circus act," Jennifer told me. "People would stop to look at us while one of the kids who's not paying attention walks smack into a post and then we're trying not to clothesline people, catching them in the middle of a long leash." At the center of the moving maypole plodded Barkley, his droopy bloodhound ears swinging, his droopy bloodhound eyes bright and amused. His matter-of-fact behavior and his pride in his work inspired Jennifer. She could walk with her eyes calmly lowered, like Barkley, focused on the task at hand. Her awareness that they were semi-ridiculous faded. And sometimes people who felt obliged to comment on her family— "You sure do have your hands full!" or "What's wrong with them?"—offered positive remarks instead, like "What a beautiful dog."

Jennifer's career as a shut-in was at an end.

Four days after moving to Atlanta with his new family, Barkley alerted Mike and Jennifer to a jailbreak. He barked, bucked, and charged about restlessly at the back door and, sure enough, Ben was gone. They released Barkley outdoors with the command: "Find Ben!" and he instinctively ran through the three phases of tracking: he roamed in wide half-circles looking for the freshest traces of Ben (Search phase), slowed down and moved several paces along an invisible path (Deciding phase), then locked in and sped up, nose down, after the boy (Decision phase).

The career of the Cobb County Neighborhood Intruder had also come to an end.

EVERY MONTH AT 4 PAWS, new families prepare to meet their dogs. Parents' hopes skyrocket. They try to help their children crank up some excitement—"Are we going to meet Cookie?" they enthuse. "We're about to meet your doggie!"

4 Paws trainers—most of them young women in khakis and polo shirts—distribute treat bags to the families, filled with goodies with which to start winning the affection and obedi-

ence of their new dogs. The parents are keyed up, full of nervous laughter. They've staked months or years of hope, logged
hundreds of miles, and raised thousands of dollars to reach
this day. Some children begin to stim—hum, or rock back and
forth, or drum with their fingers on their heads—more rapidly
now, sensitive to the heightened electricity in the room. The
air is nearly buzzing, as if the cicadas from nearby fields have
stepped inside.

In the middle room, the dogs have come out of their crates
and now line up beside their trainers. One by one, like fashion
models on a runway, beautiful dogs of all sizes will parade into
the room. Each will be led to his or her new family. But suddenly
here's Jeremy warning them against the Lassie Myth.

What? The parents weren't necessarily thinking like that,
but . . . now that he mentions it . . . okay, it rings true. Every
family in this building desperately needs a miracle.

CHAPTER 3

Connor

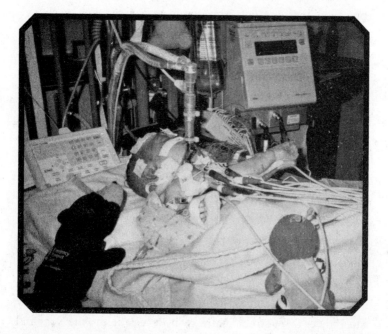

Connor Millard was born on March 1, 1999, nine weeks early, a squirming purplish almost-translucent undersea creature. He recoiled from the light and choked in the sharp silver air. A nurse scooped him into a blanket and sprinted away. After various postnatal procedures, the new parents, Deb and Scott Millard, held hands in quiet elation on Deb's hospital bed and waited for their baby to come back. In their mid-twenties, they looked like college students. Scott had a long face under a flip-top of auburn hair; his eyes held the possibility that a great punch line was just moments away. Dressed in a gray T-shirt, jeans, and sneakers, he was a pretty fit guy, an after-work jogger and weight lifter. Deb, in a cotton hospital gown, had a pale oval face, dark arched eyebrows, and a silky fall of black hair behind which she sometimes ducked and withdrew, like Violet, the shy cartoon teenager in *The Incredibles*. She was a preschool teacher. Scott was an information technology consultant in the transportation industry; he often flew to other cities to "facilitate large system implementations." In Mount Arlington, New Jersey—a hilly, wooded prewar lake resort, now a blue-collar commuter town—they were a respectful and well-liked young couple in church and village circles.

They tried to consider the possibility that something was seriously wrong with their baby, but they were ecstatic that Connor had been born alive. Doctors had warned, across a perplexing, difficult pregnancy, that the fetus might not be viable. But here he was! Alive! As he rocketed away in the arms of a nurse, he looked darling to them. Scott peeled off on a baby- or fact-finding mission. He returned to Deb's hospital room with a wistful shrug.

Four hours later, as the young couple catnapped, a nurse came for them. Despite Deb's demurrals that she could walk, the nurse steered her through cold hallways by wheelchair to the Neonatal Intensive Care Unit [NICU], a pristine hissing sanctuary of miniaturized waxy-looking infants under glass. So that's where their little fellow had fetched up! In a clean plastic bassinet as thickly wired as an early desktop computer, Connor lay beached. He was pink now, rather than purple, which seemed a promising development. His paper-thin chest puffed in and out like the cheeks of a frog. "He's incredibly cute!" Deb and Scott told each other. "His blond hair is a total kick!" A doctor arrived and explained the dilemma: Connor, though born at thirty-one weeks, had lungs resembling those of a gestational twenty-two-week-old.

The Millards, bewildered newcomers here, nodded submissively. They understood that Connor wasn't breathing on his own. *Once his lungs start working, we'll take him home and everything will be fine,* they concluded. They got more strange news later, when a different doctor explained that the baby presented "dysmorphic features"—a flat nasal bridge, turned-in fingers, skin webbing between toes, and poor skeletal alignment. *So he'll be a funny-looking little kid,* the Millards thought, and asked, "Can we take him home soon?" Not yet, they were told.

When Scott poked a finger into the bassinet and Connor grasped it with two hands like a baseball bat, the charged sweetness of contact made the new father feel hopeful. They whispered nursery songs to him, and he jerkily turned his head to gaze at them through the plastic. When nurses jabbed him with needles, he flailed around and knocked out his tubes and wires. *What spunk!* the Millards thought. *He's just an incredible little tiny guy.*

Weeks passed. One afternoon in the second month of Connor's life in the NICU, a nurse approached Deb and murmured: "You have *got* to get him out of here." And Deb suddenly saw that whatever it was Connor was "presenting" required a pediatric sophistication beyond the reach of a good regional hospital; that this everyday rhythm of Connor's gaining an ounce or losing an ounce, of new doctors stopping by to peer at the inscrutable infant over, or through, their reading glasses, of Deb and Scott rotating in and out and enthusing over a tender moment of eye contact through the Plexiglas was not advancing the boy's cause.

At eight weeks of age, Connor was transferred by ambulance

to Philadelphia Children's Hospital. Deb and Scott checked into the Ronald McDonald House nearby. On their second night in the room, they got an urgent phone call from the hospital to come quickly. They raced over to Connor's hospital room, but were denied access. Someone explained that he had "coded" and was being given an emergency tracheostomy—the installation of an endotracheal tube through his throat—and no one knew how long he'd been without oxygen. Medical personnel dashed in and out of his room, seeming to glance over at the parents with looks like, *What has New Jersey sent us?!*

When finally admitted to Connor's room, the Millards beheld a monstrosity: a swollen unidentifiable unconscious balloon-creature punctured everywhere by tubes and wires, his arms splayed out and his neck an open trach site. He was in an induced coma. The doctors spoke to them of "brain bleeds" and "brain damage." In the small hours of the morning, Scott and Deb phoned their parents in New Jersey and whispered: "We may lose Connor."

They sat up all night. A consulting neurologist broke the news to them the next morning: "Your son is going to stay in this vegetative state. He will never walk, talk, or play." There were institutions that took in infants like this, he noted. When the doctor left them alone in a small conference room, they held hands, bent over, cried, prayed, and found themselves rejecting the prognosis. "Who was that guy anyway?" they said angrily. "We never even saw him before!"

"Let's give Connor all our love, in case in some way he can process it," said Scott.

They began immediately to try to shed every hope of a normal childhood except one: that their son would know he was loved.

Finally they were permitted to hold their blank boy, an inanimate humanoid at life's margin. Deb smoothed his blond Mohawk with a baby hairbrush. In the coming days, rummaging through the diaper bag for the little hairbrush would become one of her favorite things to do—it felt so normal. She read picture books aloud in a lively voice and made toys dance outside the high-tech bassinet. "Here's Clownie! Hi, Connor! Hi, Clownie!"

"Connor's a mystery baby," she reported to Scott by phone after he returned to work in New Jersey. "A world-renowned geneticist is *very* interested in him. She came in with a yellow pad and said, 'Hmmm . . .' and she came back with a *flock* of students and they were like, 'Wow.' "

One day, as Deb bustled about near the bassinet, she glimpsed something moving across the infant's puffy face. Was it a new kind of tremor, now encroaching on the facial muscles? She looked closer. Connor seemed to be grinning. "Hi, precious, hi, Connor!" she cried. With shaking hands, she dialed Scott's cell phone. "He's waking up. He can't turn his head, it's kind of stuck to one side. And he can't bend his legs, but he's arching his back a lot." Scott flew in Thursday night and Connor seemed to kind of squint in his direction through the plastic. "He sees us! He knows us!" They rejoiced.

UNDIAGNOSED, FIVE-MONTH-OLD CONNOR WAS SENT home by ambulance. An EMT carried the baby upstairs to the apartment and a full-time nurse moved in. Under her tutelage, Deb and Scott mastered the sterile intricacies of the tracheotomy (the artificial airway in the neck), the ventilator, the oxygen tank, the heater/humidifier, the heart rate monitor, the respiratory rate monitor, the glucose monitor, the oxygen saturation monitor, the suction machine, the nebulizer, the CO_2 monitor, which had to be cross-checked against the ventilator settings, the nasogastric feeding tube, the feeding pump, the antiseizure drugs, the pulmonary disease drugs, the CADD-Prizm pump, the syringe pumps, the med ball pumps, and the IPV machine, which could be used instead of the Vest Airway Clearance System.

Scott's colleagues gave him their vacation days, but it was obvious his high-flying career was over. Bethlehem Church in Randolph, New Jersey—already praying for the Millards and cooking for the Millards—made a job offer and Scott became the church director in charge of outreach and worship.

The Millards took the baby's blood pressure, administered his IV meds, and drew blood-work into different vials with doll-sized surgical instruments. They grew conversant with the inflation line, the obturator, the flange, the Fome-Cuf, the side-port connector, the tracheal button, and the inner cannula.

Then the full-time nurse moved out and family life began!

FROM AFAR, THE APPARATUS IN the nursery looked more like a science experiment than like a crib, perhaps something involving hydroponic plants, but inside the artificial environment, a baby stayed alive. By a year old, he was a wan sitting-up boy with translucent skin, rose-colored hollows around the eyes, sparse hair, a flattened stubby nose, and a crooked smile. There was something a little uncertain or unfocused about his features, as if he'd had cleft lip repair, which he hadn't; and his eyes held a wary look, in case needles lurked nearby. But within his mechanical habitat, during breaks between medical procedures, he experienced sweet encouragement and gentle fun.

When Connor was two, the family moved to a ranch house on a blacktopped street. There were hardwood floors, a stone fireplace in the den, and a redwood deck above a small green backyard. The Millards had no knickknacks, souvenirs, or framed posters of the sort people collect on vacation—they had no life outside the health care industry—so the simple house was sparsely decorated, the painted walls mostly bare.

In this house, against all predictions, two-and-a-half-year-old Connor learned to walk. One parent crouched in front of him, arms extended, and the other parent inched forward at his side, quietly wheeling the equipment cart to which Connor was affixed by the neck. His gait was wide-legged, swaying, and unbalanced. He couldn't explore on his own farther than the eighteen inches his tubing allowed. Stiff-legged, he wobbled across the den, like an astronaut walking on the moon, using high-tech breathing equipment to survive.

Connor found his voice: a breathy squeak. High-octave little notes were tossed aloft on the whooshes of air from the breathing pump. His vocabulary was limited, his cognitive development a bit stunted, but he was emphatic. In the evening, he could sum up his day for Scott with one word: *"Mama."*

Scott asked, "What about Mama?"

Connor said: "Yeah."

And Scott kind of knew what Connor meant.

At the hospital, they'd rejected the neurologist's prognosis that Connor would never emerge from a vegetative state and they'd clung to a few words of advice from a physical therapist: "He's just a little kid. He'll be more like other kids than different from them. Just treat him like a kid." Now, when the Millards' friends signed their children up for preschool gymnastics, they signed up Connor (though it took most of the class period to move him, his gear, and his mother or father from one side of the mat to the other). He attended morning preschool and worked on little puzzles or pictures while his mother hovered nearby. He couldn't memorize more than a couple of letters but, by age four, he could dismantle and put together a ventilator circuit perfectly.

They tried a playdate with a sweet little boy from church named Christopher. The children sat near each other amid a heap of Christopher's toys and made friendly eye contact. Deb felt it was a success. As she packed up to leave, she made plans for their next visit. Christopher's mother, Tina Hardin, protested: "Good heavens! Let us come to you! It's too much work for you!" but Deb insisted. Watching Deb expertly snap together multifaceted components until a small wheeled medical unit stood in her living room and then take it all apart again an hour later, Tina murmured, "I could never do what you do."

"Oh," said Deb, "you'd be surprised what you can handle."

Connor was nearly five years old before he seemed to notice that he was an unusual sort of child. At bedtime one night, he fondled his trach and reached up with his other hand to stroke Deb's smooth neck. "Mama no?"

"That's right, Mama doesn't have a trach."

"Mama can't breathe!" he said. He seemed to have surmised that, while other people gasped and struggled for oxygen, he was a lucky boy whose parents outfitted him handsomely for life on this difficult planet.

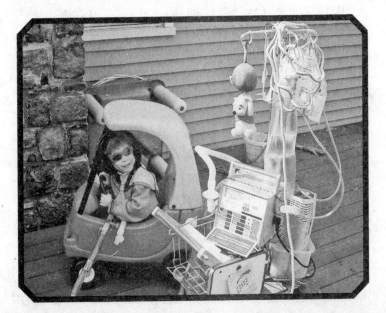

Connor's friends didn't pay much attention to his singularity either. Christopher would later say he saw Connor as a perfectly ordinary boy encased by his parents—for no apparent reason—in complicated gadgetry. When, years later, Deb ran into a teenager who'd been in Connor's gymnastics class in childhood, she asked him, "Just out of curiosity—what did you think was going on with Connor?"

The boy replied, "I thought Connor was a fireman."

BUT IT'S HARD TO STAY ON TRACK developmentally when strangers are coming at you from all directions wielding needles and scalpels, when your earliest memories are of pain, fear, immobility, insufficient oxygen, brisk handling by strangers, and the loneliness of sterile environments. At five, he began showing signs of what may have been post-traumatic stress disorder, or autism spectrum disorder, or oppositional defiant disorder, or obsessive-compulsive disorder, or some combination of psychosocial issues.

At home on the sofa, he panicked when he heard an ambulance or fire truck in the distance, or dogs barking or kids yelling down the street. He might dislike the texture of a food, or find his first taste of it to be warmer or cooler than he'd expected. The tantrums Connor threw were sometimes of such force that they seemed to alarm him as well as his parents: his eyes looked confused, as if he were a passenger in a truck jouncing down a potholed road. He hollered and writhed on the sofa until he fell asleep, and awoke with little recollection of what had infuriated him. Deb speculated that the rages were seizure-like.

Deteriorating emotionally, Connor grew terrified of squirrels, insects, and puddles. When autumn leaves drifted onto the deck, he said it was "hair" and refused to go outside. He didn't like crowds, strangers, or other children. Preschool and gymnastics were out of the question now; and then Christopher's family moved away and playdates ceased. Connor boycotted extended family gatherings. He asked only to be left alone to lie on the den sofa all day every day, wearing his favorite shirt and his favorite shorts and his favorite socks and gnawing on his anxiety-reducing chewy tube and watching his favorite shows and videos—*Blue's Clues, Clifford the Big Red Dog,* and *Thomas & Friends*—over and over and over. He was content marking off the hours by swallowing his pills in the proper order, sucking in his nebulizer treatments, and critiquing his mother's technique as she cleaned and refreshed his equipment.

From disinterest or lassitude, he nearly lost the ability to wobble-walk.

He spoke less, an occasional monosyllable every few hours. If compelled to leave the house with his mother, Connor donned dark sunglasses, a baseball hat, and headphones and chomped down on his chewy tube. Despite the preparations, the moment they exited the front door and were hit by a blast of sunlight and the horrible approach of clouds, Connor began to shake. If a bird chirped or a plane flew overhead, he came unglued. He covered an ear with one hand and frantically gestured at his mother with the other hand to take him back to the house. If she overruled him, he stiffened and refused to bend into his car seat. He choked and gagged until the ventilator alarm went off, and Deb struggled back into the house and down the basement stairs with him. On a carpeted corner of the basement, she and Scott had created a windowless lamp-lit calming environment with mats, a large rubber gym ball, and a hammock swing. It took hours of holding and rocking Connor to settle him after a disastrous attempt to go down the front steps. Gone were the days—such happy days, they seemed now!—of driving him and his medical apparatus to preschool and gymnastics.

For Scott and Deb, the imperative to keep their son alive was now joined by the desperate desire to see him *live* a little, in case (though neither spoke this thought aloud) they were nearing a steeper decline. They saw Connor's vitality draining away but couldn't find the leak. Trapped in the house by their son's agoraphobia, they felt alone and a little scared.

Could he be retreating from the outside world because he was deteriorating physically? The doctors were zeroing in on a genetic metabolic disorder, but hadn't confirmed a diagnosis or made a prediction about the length of Connor's future. Deb and Scott had to believe that, if they revived their son's spirit, his body would follow. "He used to be such a happy, engaging kid, but life scares him now," they agreed. "But how do you get a kid, literally hooked up to a machine, engaged with the world?"

Karen & Piper

hile warning clients against the Lassie Myth, Karen Shirk, like 4 Paws head trainer Jeremy Dulebohn, skips over a few personal details. The first: twenty-six years ago, a beautiful black German shepherd dog rescued her from physical incapacity, isolation, suicidal thoughts, and accidental death. The second: today, another special dog—a smaller and fluffier one—never leaves Karen's side, protecting her health and giving her happiness in uncanny ways.

These days, even after long and occasionally recurring sieges of illness and invalidism, Karen—stocky, dressed in bulky blue jeans, a man's white T-shirt, and a camouflage-colored billed cap, swaying deeply as she walks, occasionally motoring forth in a wheelchair, breathing through the metal tube in her throat—retains traces of a long-ago lighthearted girl. Her swingy light-brown hair falls in the same pixie cut she wore as a ten-year-old, and her round face and cheeks, mischievous expressions, and willingness to laugh long and hard are also throwbacks to the kid version of herself. Ditto the endless delight she takes in the dogs, in all animals really. So, when she's obliged by poor health to huff and shuffle down the cement hallway of her dog-training school, the effect is that of an impish child in adult clothes pretending to limp and lean on a cane.

Her windowless, wood-paneled office sits at the far end of the one-story dog-training school. Crammed with massive wooden desks, metal filing cabinets, refrigerators, and tables, the office seems to have been mistaken for a storeroom. Manila folders rise from every horizontal surface, imprinted with fading brown ripples of coffee-cup circles. Dog calendars, framed photographs of dogs, laminated dog stories from newspapers, and children's crayoned stick figures of dogs have been taped up pell-mell all over the walls. Dog stuffed-animals, tiny glass dogs, and kitschy ceramic dogs crowd the bookshelves, while coffee mugs with dog pictures on them huddle around the coffee machine, all gifts from grateful families.

A real dog pack lives here, too.

Every time Karen steps into her office—even if she's been gone only fifteen minutes to peek in on a class, check on puppies in the nursery, meet a job applicant, or warm up her coffee—a dozen sassy little dogs rejoice as if they'd feared never seeing her again. It's like trying to cross a field popping with grass-hoppers. Little dogs come bursting all at once out of their cave under the central wooden desk and from hiding places under other pieces of furniture. These are the papillons, silky toy spaniels whose name is the French word for "butterfly" and refers to their wide upright fringed ears like butterfly wings. At the Return of Karen, some spin in excitement; others take turns leaping daringly onto Karen's black leather desk chair and back to the floor; and a feminine lass with a froufrou tail is tap-dancing across the computer keyboard. All raise tiny pointed faces and emit high-pitched yips of hallelujah.

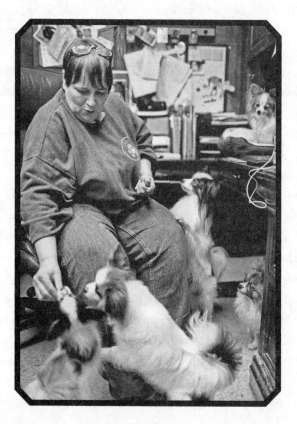

Karen casually brushes dogs off her desk and chair in order to seat herself. The small animals settle at her feet, collapse their ears like umbrellas, and stare up at her adoringly. One black-and-white boy named Pirate is unable to dial it back; he vaults to Karen's shoulder and clings there wide-eyed as a lemur. Karen detaches him, returns him to the floor, and urges him to calm down. Under the collective gaze of the pack, Karen wheels her chair around and begins replying to emails and taking phone calls. Whenever she chuckles, the little dogs take out their ears like kites and wave them around.

The special dog of Karen's middle years is a papillon, which surprised her. She wouldn't have thought a silky lapdog would be her type. She dresses plainly, speaks plainly, and lives with her four school-age children (one adopted from Bulgaria, three adopted from Haiti) in a spartan farmhouse surrounded by fields and a horse barn. But a few years ago, an Ohio breeder of papillons phoned 4 Paws and offered to donate a puppy born with three legs. If Karen couldn't use the puppy, it might have to be put down. "Great, just what I need, a three-legged papillon," she'd laughed, but had accepted the puppy to save its life.

"Gracie" delicately hopped her way through the building, sniffing and exploring. One day a staff member noticed that Gracie was picking up commands. When the dogs in the training circle in the social hall were told, "Sit," Gracie stopped whatever she was doing, wherever she was, and sat.

"Are all papillons this smart?" Karen mused. She checked out a popular list of the most intelligent breeds. Based on the speed with which purebred dogs mastered, obeyed, and remembered new commands, Dr. Stanley Coren, a professor of psychology at the University of British Columbia, had compiled a Top Ten list of the smartest. Karen (while appropriately skeptical that such a

list made sense when applied to an individual dog) was amazed
to read the following:

Border collie

Poodle

German shepherd

Golden retriever

Doberman pinscher

Shetland sheepdog

Labrador retriever

Papillon

Rottweiler

Australian cattle dog

Papillon? She nearly spit out her coffee.

The papillon breeder offered another puppy to Karen: Gracie's
sister, Piper, with a white-and-auburn coat. From the first, Piper
was enthralled by Karen, mesmerized by Karen, in love with
every single thing about Karen; clearly she was a one-person dog
(not all dogs are) and *she claimed Karen.*

Piper occasionally leaped into Karen's lap in a strange frenzy,
pawing at Karen's chest and licking her face. "Stop, stop," Karen
laughed, pushing her away, not always finding it a convenient
moment for a love-fest. Sometimes she was on a phone call
when Piper did this; sometimes she was meeting with a parent or employee. Just shy of being annoyed, Karen had a hunch.
She began jotting down the times of Piper's frantic greetings. It
showed up immediately that, in every case, Karen's chronic neuromuscular disease flared up later that day, sometimes within the
hour. Piper's occasional enthusiasms were, in effect, untrained
"alerts" of approaching episodes of acute respiratory distress that
gave Karen time to alter her schedule and prepare medications.

In time, Karen would learn that about 10 percent of all dogs alert naturally to unhealthy physiological changes in a beloved human, and that, among papillons, the rate may be 50 percent. As increasing numbers of families whose children had seizure disorder or diabetes applied to 4 Paws, papillons grew in demand. How does the staff know which papillons have the gift? Almost invariably, they identify themselves by warning someone in their foster family or on the staff that some kind of physiological discomfort is imminent; even if one fails to self-identify, Jeremy can usually spot a pup with the gift. Nearly all Karen's papillons are on track to become service dogs. They're not the right match for rough-and-tumble kids (a papillon couldn't offer much help to Logan Erickson), but they're perfect for older children with deafness, seizure disorder, or diabetes who can be gentle with the little animals.

Karen steered Piper into medical-alert and obedience training. When Piper graduated, Karen ordered a red 4 Paws for Ability service dog vest, size extra-small. She would no longer be caught unawares by a sudden inability to breathe. And now, wherever she went—into a restaurant, onto a plane—her tiniest friend could come, too.

NOW TEN YEARS OLD, PIPER remains Karen's *Numero Uno*. At 4 Paws, she holds the place of honor: a plush pink bed on Karen's desk, to the left of the computer monitor.

"Who's so pretty? Who's so pretty like Marilyn Monroe?" cries Karen in a high-pitched voice. Piper sits up tall to perform her Marilyn Monroe: she tosses back her long fringed satin ears and strikes a pose with her white-striped muzzle high in the air. "Who's so pretty like Marilyn Monroe?" exclaims Karen again and Piper arches her back and holds up her snoot even higher.

"Piper," says Karen in a lower voice, testing her. "You want a bath? Piper want a bath?"

Piper drops the Marilyn Monroe act at once. The happiness falls from her face as she backs up into the darkness behind the computer. She presses flat against the wall, trying to elude the reach of Karen's fingers.

"No, I'm teasing you, Piper! I'm sorry!" says Karen. She lifts her voice into baby talk again: "Piper: *Chipotle*? You want to go *Chipotle*? You want to go Sam's Club or you want to go Chipotle? You want to go Chipotle and *eat*?" Piper leaves the back wall, struts out onto the desk as if it were a balcony. She swishes her feathery tail at the mention of Sam's Club, but the idea of Chipotle makes her absolutely delirious. She cocks her head, listening. "*Chipotle*? Go *bye-bye*? You want to go? You want to go *Chipotle*? Go Chipotle bye-bye?!"

Piper bounces backward on all fours, twice. Chipotle is the Mistress's and Piper's favorite spot for lunch! Now she makes a quick circle on the desk, like a debutante darting around her bedroom to gather lipstick, a clutch purse, and a light wrap. She soars gracefully to the floor, dodges the papillon hoi polloi, and dashes to the office door, where she sits and waits to be helped on with her tiny red service dog vest.

In the booth at Chipotle, beside Karen and across the table from Jeremy, Piper trembles all over with delicious expectations, which are quickly gratified. The Grilled Chicken Burrito Bowl is the Mistress's and Piper's favorite. "I know! Don't look at me! I'm the worst service dog owner in the world," moans Karen while handing over a rolled-up ball of rice and chicken to Piper.

"This is true," says Jeremy, who instructs clients never to feed table food to their dogs because it quickly becomes a bad habit impossible to break. "But . . . you're also my boss, so it kind of balances out."

Back at 4 Paws, Piper frisks into Karen's office, pauses for the removal of her vest, maneuvers her way through the doggie crowd, sails up to her penthouse, and closes her long-lashed eyes for her beauty sleep.

MOST OF THE PAPILLONS SEEM CONTENT with their collective life of yappy shenanigans under the gaze of their indulgent, chuckling deity. But Abel, a feisty tricolor male (black-and-white all over with a few touches of tan on his face), can't *stand* Piper's favorite-child status. Sometimes, after Karen and Piper have left for lunch, Abel bounces up to Piper's pink bed, pees on it, hops back down, and hides in the cave under the second desk. One pictures him chortling, his small shoulders going up and down. One time he pooped in Piper's bed! Score!

Occasionally his challenges occur in Piper's presence. Having returned from her lunch date, she dallies on the floor, catching up with old friends and relations. While her back is turned, Abel vaults up to her pretty pink bed and stands in it, wide-legged and cocky, showing off for his homeboys under the other desk. One pictures *their* shoulders going up and down mirthfully.

"I always know what Piper's thinking," Karen says. "I'm not an 'animal communicator.' Ha! I'm not a 'psychic.' No way! I've met those people! They're like, *I'm seeing something that looks like . . . a can of dog food.* And I'm like, 'Oh really? What was your first clue?' I am not 'reading Piper's mind.' But I know what she's thinking. I'll be sitting at my desk typing and I'll feel an intense gaze boring into me from the floor. I look down and it's Piper. I know perfectly well that she's asking me: *Are you* seeing *this? Are you planning to allow this to* continue?! I don't even have to look over to know Abel's in her bed."

Without turning around, Karen calmly says: "Abel, down." The little fellow seems to exhale peevishly, with a kind of angry dish-towel whip of the ears, before dropping to the floor and slinking back into the desk-cave.

"AN ANTHROPOMORPHIC TRAP LIES IN WAIT for all who study dogs," warns Dr. Ray Coppinger, professor emeritus of biology at Hampshire College. Reviewing a book by a fellow dog expert, he notes that the author "is at his anthropomorphic worst" in describing dogs who seem to think and feel in completely human ways: one dog displays "rapt attention," another "looks questioningly," and a third dog whines "in a pleading manner."

The problem for a writer is that anthropomorphic descriptions bring dogs to life on the page far more engagingly than if Coppinger's author-under-review had reported, of those three episodes: "The dog looked," "The dog looked," "The dog emitted a sound." But Dr. Coppinger was looking for science, not fiction.

It is, in part, a storytelling problem. The British novelist E. M. Forster famously wrote: *"The king died and then the queen died* is a story. *The king died, and then the queen died of grief* is a plot."* The former reports an event in time; the latter relates the event with an underpinning of psychology and conflict.

Thus (from the book Coppinger reviewed), a story: *The dog whines.*

A plot: *The dog whines in a pleading manner.*

And . . . a story: *While Karen's back is turned, Abel jumps up to Piper's bed and stands in it.*

A plot: *While Karen's back is turned, Abel vaults up to Piper's pretty pink bed and stands in it, wide-legged and cocky, showing off for his homeboys under the other desk.* The latter, in Dr. Coppinger's terms, has clearly been written by a writer at her anthropomorphic worst.

And yet, these days, the frontier—between objective reporting and anthropomorphism—is on the move. Observations that seem to blatantly attribute human thoughts and feelings to animals are being taken apart and examined plank by plank by cognitive researchers. When their results are published in peer-reviewed journals, the bright lines dividing humans from nonhuman animals are sometimes erased.

DOES PIPER LOVE KAREN? Is Abel jealous? *Do* animals think and feel, or is this anthropomorphic nonsense?

In the mid-twentieth century, while animals remained *personae non gratae* in the upper echelons of Psychology and Behaviorism, a few species began to be stalked in their natural habitats by great *field* scientists. Mud-spattered and booted, these wildlife biologists with their notebooks and helmets, binoculars and insect repellent, plunged into rain forests and through swamps, tramped up mountains and across glaciers, crouched in silence for days, or *years,* drew their own conclusions and penned their own masterful bodies of work. They perceived animal worlds crackling with the intensity of family feeling and complex social relationships.

The preeminent Western field scientists include three Nobel Laureates credited with co-founding the field of Ethology, the study of animal behavior: Dutch-born British researcher Nikolaas Tinbergen (seagulls), Konrad Lorenz of Austria (jackdaws, graylag geese, ducklings—he discovered "imprinting"), and Karl von Frisch of Austria (fish and bees). Other greats include Mardy Murie (the "grandmother of the conservation movement," Arctic wildlife), George Schaller (mountain gorillas, lions, giant pandas, snow leopards), Dian Fossey (mountain gorillas), Jane Goodall (chimpanzees), Hans Hass (stingrays, sharks, and animals of the coral reef), Sylvia Earle (oceanography), Frans de Waal (bonobos, chimpanzees), L. David Mech and Luigi Boitani (wolves), Cynthia Moss, Iain Douglas-Hamilton, and Joyce Poole (elephants), Nigel Franks (ants), Alan Root (wildebeest, animals of the Galapagos),

E. O. Wilson (ants, then an encyclopedia of life), Bernd Heinrich (moths, bees, beetles, flies, dragonflies, butterflies, and ravens), and the novelist and nature writer Peter Matthiessen (snow leopards, shore birds, tigers, and cranes). Local experts worked beside them as mentors and teachers, guides and assistants, translators and drivers, their names mostly unknown as they tended not to publish the results of their field-work or host film documentaries.

Together, the field reseachers found personality and intelligence to be the foundation of everything they observed. Though embraced by the public, many of these pioneers were shunned by their counterparts in universities and research institutions (still under the sway of Behaviorism) for attributing invisible mental processes like thoughts and feelings to *animals*.

THEN CAME THE "COGNITIVE REVOLUTION," an intellectual pushback against Behaviorist dogma. "Behaviorism was an exciting adventure for experimental psychology," writes Professor George A. Miller of Princeton, "but by the mid-1950s it had become apparent that it could not succeed." The missing piece was thinking, cognition, "mind." The stimulus/response model fell short as an explanation for the rich complexities of human behavior. *Why should human mental processes be deemed nonexistent just because they're invisible?* researchers wondered. Some invoked a time-honored observation: "Absence of evidence is not evidence of absence."

With the rise of neuroscience and the invention of extraordinary technologies for peering inside living brains, respect for *thinking* as an activity worthy of scientific scrutiny was restored, with the tantalizing dream that consciousness itself might someday be netted and defined. Even *emotion* began to be respected as a component of thought rather than dismissed as a sentimental footnote to pure reason.

"THINKING" CERTAINLY SEEMS LIKE AN ordinary, everyday kind of thing, but it has remained mysterious even under the klieg lights of the new technologies. "Exactly *how* the physical brain produces nonphysical consciousness remains a great mystery," writes neuropsychologist Dr. Rick Hanson, who is affiliated with University of California, Berkeley. Broadly stated, thinking is the brain-based activity of taking in and processing information, during which ideas, words, feelings, images, fantasies, music, memories, intentions, interpretations, expectations, opinions, judgments, beliefs, and plans may be generated.

Do dogs think?

"Maybe yes, and maybe no," writes Dr. Ray Coppinger. "What dogs do quite well, though, is make people think that dogs can think."

Dogs do *appear* to think. Is that worth anything?

My son Lee and his girlfriend, Maya, recently watched as their gray, curly-haired, two-year-old schnauzer-mix pup, Charlie—who, despite a quite impressive mustache and beard, looks and acts like a little bright-eyed boy—hopped off the sofa with a bone in his mouth and headed to the bedroom, there to gnaw his bone in privacy. He accidentally dropped the bone, and trotted proudly down the hall without it. Halfway down the hall, Charlie paused, and looked back. He stood there uncertainly. Then he continued down the hall. He stopped again, looked back a second time, turned around, scampered back to the sofa, picked up his bone from the floor, and *then* trotted proudly down the hall.

A minor moment in life-with-dog, of course, the sort of moment millions of people observe every day with their millions of dogs. And yet another Charlie, the greatest biological observer of all time, made a little something of it when he observed similar behavior. "[Charles] Darwin used observations of behavior such as pausing before solving problems to support his claim that even animals without language are able to reason," writes Mark Bekoff, professor emeritus of ecology and evolutionary biology at the University of Colorado in Boulder. Darwin, as *The New York Times* has noted, "was comfortable with the assertion that animals have thoughts, plans, and feelings."

That stance won him roughly as many followers as his theory of evolution.

CHARLIE THE DOG SPENDS MOST WEEKDAYS at our house with me and my three dogs while Lee and Maya are at work. The youngest of our family's pack is Henry, a scruffy, wire-haired yellow one-year-old terrier mix. Henry loves Charlie, but has one complaint: every day, as we come out of the woods behind the house after a walk, Charlie runs ahead up the dirt path, hides, and waits, planning an ambush. As Henry comes blithely up the path, Charlie tackles him. Henry squeaks and rolls into a ball as Charlie wrestles him to surrender.

Henry loves Charlie, but hates this game.

Charlie loves Henry and finds that this game never fails to amuse.

One day last month, Henry outwitted his stalker. As usual, Charlie ran ahead to conceal himself for the hunt. Henry slowed down, and then stopped. That was the moment that a new thought may have been generated. Suddenly he *revved up* and bounded leftward off the path! He bushwhacked uphill into our next-door neighbor's backyard, ran the length of it, took a sharp right, snuggled under their decorative wooden fence back to our side, and then sprinted for our back steps and safety. When Charlie, in the woods, gave up waiting for Henry and headed home, mystified by the disappearance of his prey, there stood Henry on the steps, having arrived there first. Henry wore a bemused expression, like *Did I really just do that?*

Henry deployed his new strategy every day that week, by the end of which he was simply beaming from the back steps when Charlie came into the yard. He was tickled with himself. But then Charlie wised up. Though I didn't witness *his* moment of introspection, I saw the results: he stopped running so far ahead of Henry to prepare the surprise attack; instead, he stayed about four feet ahead of Henry and looked back frequently. I had to tempt Charlie with treats to move him up into our yard, allowing Henry a chance to make his escape.

Last Tuesday, it was Henry's turn to do some more thinking. His best friend and nemesis was spending the day at home rather than with us. Charlie was *not* with us as we tramped into the woods, nor when we turned about and headed for home. According to his new habit, Henry deliberately fell behind the other dogs on the return route, preparing to make his leftward end run to the neighbor's backyard to avoid Charlie . . . who wasn't . . .

with us. I glanced back and saw Henry standing at his usual fork in the path, looking thoughtful again. Suddenly he *blew* by me, running safely and joyfully up our own path into our backyard, ears flung back and mouth wide open as if in laughter.

It's difficult for us to know how dogs or any other animals think. We humans are so pleased with our own word-saturated mental processes that non-language-based thoughts strike us as inefficient. Are they like picture books, flip books, or silent movies? Are they odorous or kinetic? Maybe they're like the diagrams of MapMyRun.com, or like the street views on Google Maps. Are bird thoughts like Google Maps' satellite views? Maybe migrating birds and homing pigeons have minds that resemble the aurora borealis: the string music of the magnetic spheres.

Whatever they are, animal thoughts strike us as not quite *comme il faut*. But it's hard to explain Henry's actions without a tip of the hat to something going on inside that spiky-haired little head: *Charlie's not here today. I can safely run straight up our yard to the back steps.*

The following day, a Wednesday, Charlie rejoined us. As our daily walk ended, I lured the little curly gray boy, with treats, into a last romp in our backyard, and glimpsed a wire-haired yellow blur zooming up our neighbor's driveway.

BUT "THINKING," SINCE ANCIENT GREECE, has been celebrated as an exclusively human achievement, and "mind" has been idealized as a disembodied, immortal, divinely touched sphere that exists apart from—and is nobler than—the human body. How could animals trespass on such sacred ground by having minds of their own?

As a new generation of neuroscientists began looking for qualities like intelligence, personality, and even morality within the moist folds of human brain tissue, Behaviorists scoffed and spiritually inclined people were offended, as if such research ruled out any possibility of the transcendent. And if the public felt alarmed by science's attempts to reduce the human "mind" and "spirit" to one bodily organ consisting of "three pounds of tofu-like tissue containing 1.1 trillion cells," the notion that *animal brains* were anything like our own would have confused folks even more.

But they needn't have worried: the first wave of scientists of the Cognitive Revolution dismissed the concept of "animal minds" just as the Behaviorists had. "Animal thinking" was considered a subject, noted *New York Times*, "that belonged far outside the realm of scientific exploration."

THEN, IN THE 1970S, Dr. Donald R. Griffin, a Harvard-educated professor of zoology at Harvard and at Cornell and the widely respected codiscoverer of echolocation in bats, expressed his conviction that animals might have the capacity to think and to reason, too. Weren't animals part of the same deep evolutionary history that produced the massive brains and infinite insights of *homo sapiens*? "Wasn't it possible," as Margaret Talbot wrote in *The New Yorker,* "that a chimpanzee who scoured the rain forest for a chunk of granite, then used it to crack open a nut, was consciously thinking about that tasty morsel inside, rather than executing rote movements?" Academia was scandalized! Only Dr. Griffin's tenured position and international reputation allowed him to hang on to his job. Unfazed, Dr. Griffin would publish books like *The Question of Animal Awareness* (1976), *Animal Thinking* (1985), and *Animal Minds: Beyond Cognition to Consciousness* (2001). According to *The New York Times:* "Many scientists say the only reason that animal thinking was given consideration at all was that Dr. Griffin suggested it."

Despite warnings from colleagues that they risked sidelining their careers, other researchers picked up Dr. Griffin's openness to evidence of intelligence in nonhuman animals, and not intelligence but intelligence alone, tempered by emotion.

THE TWENTIETH-CENTURY SEARCH FOR *EXTRATERRESTRIAL* intelligence has not yet borne fruit, but the scientific quest for intelligence in our fellow earthlings is yielding incredible results. Among the beautiful mysteries of the cosmos opening to twentieth-century inquiry are the minds and hearts of animals. "Only humans have human minds," writes ecologist and nature journalist Carl Safina. "But believing that only humans have minds is like believing that because only humans have human skeletons, only humans have skeletons."

Under the banner of new academic specialties like animal cognition, animal personality, anthrozoology (the study of human and animal relationships), ethology, (the study of animal behavior), cognitive ethology (the study of animal thought processes), evolutionary biology, human-animal interaction, and veterinary ethics, scientists are finding evidence of nonhuman cognition all over the place.

"A growing body of evidence shows that we have grossly underestimated both the scope and the scale of animal intelligence," writes Frans de Waal, Emory University professor of psychology, primatologist, and director of the Living Links Center at the Yerkes National Primate Research Center. "Can an octopus use tools? Do chimpanzees have a sense of fairness? Can birds guess what others know? Do rats feel empathy for their friends? Just a few decades ago we would have answered 'no' to all such questions. Now we're not so sure . . . [Recent] findings have started to upend a view of humankind's unique place in the universe that dates back at least to ancient Greece."

In 2012, an international conference of academic luminaries gathered in Cambridge, England, to publicly correct the widespread scientific misconception that animals are not "sentient"—thinking and feeling—creatures. Their pathbreaking "Cambridge Declaration on Consciousness" declared that non-human animals possess "the neuroanatomical, neurochemical, and neurophysiological substrates of conscious states"—in other words, all mammals, as well as many other vertebrates and some invertebrates, have the internal apparatus with which to generate higher-order cognitive functions, including consciousness, self-awareness, and thinking.

"BIRDS APPEAR TO OFFER," noted the Cambridge Declaration signatories, " . . . a striking case of parallel evolution of consciousness. Evidence of near human-like levels of consciousness has been most dramatically observed in African grey parrots."

The intelligence of parrots was exemplified by the talking wonder, Alex, an African gray parrot, the feathered colleague of Brandeis and Harvard University professor Irene Pepperberg, who had a working vocabulary of a thousand words (the bird, not the professor). Birds in the crow family, Corvidae, display sharp problem-solving skills, in and out of the lab. My daughter Molly Samuel, a public radio science reporter and amateur birder, attended a guided bird walk in Golden Gate Park, San Francisco. The Audubon Society volunteer began by warning the group about the local Steller's jays. Sprightly and large, with sapphire-blue bodies and tails and coal-black crested heads, they're foragers and nest robbers: they eat the eggs and the young of other bird species. The ornithologist said she'd realized, as she counted nests in the park, that Steller's jays were watching her and using her movements as clues to the treasures, after which they devoured the eggs and the baby birds of other species in the nests she'd just logged. She was willing to direct the birders to the locations of nests, she told Molly's group, but she asked them to conceal their discoveries and to bluff interest in other trees, in order to confuse the Steller's jays.

CEPHALOPODS—OCTOPUSES, SQUIDS, AND CUTTLEFISH—are knocking the *socks* off the tests. Octopuses are capable of attentiveness and decision-making, stated the Cambridge Declaration. They're creative tool-users and they're "social learners": one octopus can learn by observing another octopus. In one experiment, a freshly caught octopus is set loose in a tank that contains a clear Plexiglas puzzle-box; inside the puzzle-box swims a live shrimp, a favorite food of octopuses. The newly captured octopus huddles behind a rock and dares not approach. The puzzle-box is removed and placed in a second tank next to the first. Through the glass walls of the side-by-side tanks, the first octopus watches a more experienced octopus grab the puzzle-box, unscrew its hatch, reach in, and gobble up the shrimp. The puzzle-box, with a fresh shrimp, is now returned to the first octopus, the novice. This time, the new octopus leaps onto the puzzle-box, masters it instantly, and eats the shrimp. The California Academy of Sciences calls the octopus "the Einstein of Invertebrates." One documentary film reporter rather ominously commented: "There may be no holding back its formidable intelligence."

Fish have their enthusiasts! According to Professor Culum Brown of Macquarie University, Sydney, Australia, fish develop cultural traditions, cooperate with one another, recognize themselves and one another, and feel pain.

"Honeybees . . . can count, categorize similar objects, understand 'same' and 'different,' and differentiate between shapes

that are symmetrical and asymmetrical," according to ento-
mologists. Crickets, water striders, and fruit flies "have all been
found to have personalities—meaning that some are bolder in
their actions than others."

"Every time someone declares that they've found *the* skill
separating 'us' from 'them,' " writes science journalist Virginia
Morell, "someone else surfaces to say they've just found that
ability in another species."

SO ANIMAL COGNITION—"MIND"—IS INCREASINGLY acknowl-
edged by scientists as evident across a wide swath of animal species. What about emotion, "heart"?

No mind, no heart had been the credo of the European philosophers who believed that emotions had been designed by the
Creator exclusively for mankind. Again, Darwin saw what others
missed, including parallel expressions on the faces of human and
non-human animals. He called the movement of facial muscles
"the language of the emotions," noting that, like humans, some
animals pursed their lips when concentrating, squinted their
eyes when accessing memory, bared their teeth when enraged,
and relaxed their jaw muscles when listening closely.

That thread of inquiry was dropped for a long time. But following the Cognitive Revolution, *emotion* began to be perceived
as a component of thought rather than dismissed as a sentimental footnote to pure reason. Researchers began to look at animals
in this light.

"Human emotions, while possibly unique in some respects,
have evolved from those of mammals, which in turn have evolved
from those of reptiles, and so on," writes Dr. John Bradshaw,
director of the Anthrozoology Institute at the University of Bristol. "If emotions are . . . survival mechanisms, then they most

likely evolved to fulfill specific functions. And those functions—avoiding danger, counteracting threats, forming pair-bonds that enhance the survival of offspring—are not unique to man."

Today, investigation of animal "subjectivity" is no longer laughed out of the laboratory. Far from it: marvelous experiments are being devised, new cadres of field scientists are stepping into the wild, and marvelous books have been published. The results describe a planet teeming with sentient beings.

In a recent summary of where we stand now, science journalist and essayist Charles Siebert writes: "Advanced neurological and genetic research has shown that animals like chimpanzees, orcas, and elephants possess self-awareness, self-determination and a sense of both the past and future. They have their own distinct languages, complex social interactions and tool use. They grieve and empathize and pass knowledge from one generation to the next. The very same attributes, in other words, that we once believed distinguished us from other animals."

"PRIMORDIAL EMOTIONS"— six states of Pain, Hunger, Thirst, Fatigue, Stress, and Sexual Attraction—indisputably appear throughout the animal world. They are the engines of survival and reproduction, moving creatures of every stripe toward oxygen, water, and nutrition; safety and rest; mates and offspring. We humans experience these emotions-in-the-raw like everyone else. We just make a bigger to-do about every one of them.

"Classic emotions"— by most counts, there are six of those, too—are Happiness, Surprise, Fear, Anger, Disgust, and Sadness. They are universal among vertebrates and lately are thought to be shared by octopuses, those Einsteins of the Invertebrates.

The classic emotion, "Happiness," has been spotted by researchers in many species. "Of course dogs have feelings, and we have no trouble acknowledging most of them," writes Dr. Jeffrey Moussaieff Masson, a bestselling author and animal rights activist. "Joy, for example. Can anything be as joyous as a dog? Bounding ahead, crashing into the bushes while out on a walk, happy, happy, happy. Conversely, can anything be as disappointed as a dog when you say, 'No, we are not going for a walk'? Down he flops onto the floor, his ears fall, he looks up, showing the whites of his eyes, with a look of utter dejection. Pure joy, pure disappointment . . . The words used to describe the emotion may be wrong, our vocabulary imprecise, the analogy imperfect, but there is also some deep similarity that escapes nobody."

"Dolphins chuckle when they are happy," reports Dr. Marc Bekoff of the University of Colorado. "There's also solid scientific information that dogs laugh—there's what researcher Patricia Simonet calls 'a breathy, pronounced, forced exhalation' heard when dogs are excited and when they play . . . Rats also chirp with joy." A farm animal authority reports that "even hens love to play, and they're smart, moody, emotional, and form close friendships."

So, it's agreed: many animal species experience primordial emotions and classic emotions. Do any nonhumans feel "complex emotions," those unpredictable leaps and twists of the soul without which human life would be a dull trudge from bed to table to work to table to bed, devoid of symphonies and novels, cathedrals and movies, love affairs and tragedy, Valentine's Day and Facebook? Human emotions come in an infinite range of fine permutations, because few things are as interesting and important to people as our feelings.

Unpleasant complex emotions include Annoyance, Anxiety, Bitterness, Boredom, Chagrin, Contempt, Depression, Despair, Disappointment, Distrust, Embarrassment, Envy, Frustration,

Grief, Guilt, Hatred, Jealousy, Loneliness, Loathing, Lust, Nervousness, Outrage, Pessimism, Regret, Remorse, Shame, Suspicion, and Worry.

Pleasant complex emotions include Awe, Compassion, Confidence, Contentment, Courage, Curiosity, Desire, Determination, Empathy, Faith, Flirtatiousness, Forgiveness, Gratitude, Hope, Honor, Humor, Optimism, Pity, Pride, Nostalgia, Relief, Self-Confidence, Sympathy, Tenderness, Trust, Wonder, and Zest.

Also Love. Which, oddly, does not appear on experts' lists of primordial or classic emotions.

Do non-human animals experience complex emotions?

"Scientists generally agree that [human] consciousness is much more complex than that of other mammals," writes Dr. Bradshaw. "Indeed, we humans are able not only to experience emotions but also to examine them dispassionately, to ask ourselves questions such as 'Why was I so anxious last week?'" Most non-human animals do not appear to operate at this level of self-awareness: "All of the available evidence suggests that their emotional reactions are confined to events in the here-and-now and involve little, if any, retrospection."

Living in the Now, a spectacularly popular concept among twenty-first century humans, rules out complex emotions like Regret and Remorse, which require second-guessing onself and wishing to re-do past actions. Nostalgia, Pessimism, and Optimism also demand imaginative time-travel between the past and future. Awe and Wonder depend upon a sense of the ineffable. Bitterness calls for the long nursing of a grievance. If you live in the Now, you're not zigzagging back and forth across your life's timeline so much, nor trotting up and down stairways to heaven.

But that still leaves a sparkly collection of emotions for animals to sample. We humans are so proud of our brains, endowed as they are with massively complex neo-cortexes (neo = new, the last part of the brain to evolve). With them, we churn through a zillion bytes of encoded facts, fantasies, memories, and alternative scenarios to come up with nuanced, richly-hued, thickly-layered tapestries of thought and feeling. But all mammals have neo-cortexes. And recent studies demonstrate that the brains of animals without neo-cortexes, like birds, reptiles, and our new friends, the octopuses, also perform neo-cortex-like cognitive functions.

DO *DOGS* EXPERIENCE COMPLEX EMOTIONS? Does Juke love Logan Erickson? Does Barkley love Ben and Sam Schwenker? Are love and jealousy in play in the papillon micro-habitat of Karen Shirk's office?

Those answers would arrive a bit late, for even as the field of ethology gained traction in the last quarter of the twentieth century, inviting all sorts of creatures into the big tent of Animals Who Think and Feel, *dogs* were excluded. *No Dogs Allowed.* Dogs struck researchers as of dubious provenance: not quite wild animals, kind of man-made, and unworthy of serious study.

In 1959, when George Schaller first hiked into central Africa to live among mountain gorillas, he bushwhacked a path followed by hundreds of field scientists in years to come. But no one hiked anywhere in the hopes of hunching in the underbrush and spying on *dogs* in *their* natural habitat. That's because dogs were presumed to be *without* a natural habitat, without a biological niche. How could you study such a creature: Hide behind the living room drapes? Crawl under your own blankets at night with a flashlight?

What people vaguely thought they knew about the origin of dogs was that, sometime after diverging from prehistoric wolf ancestors, they'd been remixed and concocted—hadn't they?—by the leisure classes of Victorian England. It seemed best to leave the study of dogs to the historians of nineteenth-century Great Britain, because neither field scientists nor laboratory scientists wanted to get near them.

CHAPTER 5

Lucy

Their dreams weren't extravagant. But they were unattainable. James and Eleanor Keith, who lived in Alabama, wanted a family. They couldn't conceive a child the old-fashioned way, nor after years of increasingly sophisticated medical interventions. "For three and a half years, we were disappointed every thirty days," says Eleanor, who has a wide, freckled face, flyaway brown hair, speckled light brown eyes, and an easy smile. She comes from small-town Iowa and James from upstate New York; she worked in the central office of a national cosmetics company and James was an automotive designer. By 2004, years of hope, financial outlays, and medical procedures rested upon a microscopic cluster of cells more delicate than soap bubbles. The doctor recommended against implanting the embryo. He didn't think it would "take." But Eleanor felt herself becoming that person around whom voices dropped when she entered a room. News of someone else's pregnancy or new baby did disorient her, but it felt worse to be excluded, as if she were forbidden to be in the presence of sacred conversations about conception, birth, and motherhood. "So we paid ten thousand dollars for a twenty percent chance of success on the embryo." The Keiths edged close to parenthood for several days after the implantation, their semi-hypothetical child drifting in life's antechamber. Eleanor secretly liked the name "Lucy." But the little bubble couldn't find a grip and slipped away.

Eleanor had grown up surrounded by grimly determined, sometimes depressed people. In time, she would lose a step-niece, a cousin, and her beloved sister to suicide. Ground down by the long march of infertility, James began slipping in and out of depression, too. He often wore a pained, brave expression, squinting rather than smiling at the obligatory occasions. He was a good man, soft-spoken and capable, but life had been hard. Eleanor was an outspoken political liberal, a churchgoer, a musician, and a snow skier. In subdued family gatherings, she sat up straighter, bit her lower lip momentarily, then publicly took the bright view because *someone* had to like *something*. She "put on a happy face," which was appropriate given her profession in cosmetics management.

In 2005 they turned to adoption and on April 26, 2006, five years into the Keiths' marathon toward becoming parents, an adoption agency matched them with a ten-month-old baby girl named Mei Ling, living in an orphanage in Hubei Province,

China. "THE STORK HAS LANDED!!!!" Eleanor posted on her website. She'd found Lucy.

WHEN THE GREAT TWENTIETH-CENTURY BRITISH psychiatrist John Bowlby began to look closely at the bond between parents and children, it was the *absence* of something that intrigued him. Immersed in tragic case studies of infants and young children separated from their parents during the Second World War, observing firsthand the hollow-eyed children silently rocking themselves in orphanages, hospital wards, and refugee camps, Bowlby asked, in effect: In these delayed, stunted, and joyless children, what magic elixir is missing, without which they cannot grow and thrive?

Like a physicist theorizing the existence of an invisible phenomenon—say, dark matter—by observing the power it wields in the visible world, Bowlby reasoned his way toward a vital life force by noting that the lack of it left children at death's door.

A deeply compassionate man with a high balding forehead and enthusiastically up-angled eyebrows, Bowlby charted the stages of the physical and emotional collapse of children after their parents disappeared. In time, he would name the stages: "Protest," "Despair," and "Detachment."

His foundational *Attachment and Loss* trilogy ultimately filled three volumes: *Attachment* (1969); *Separation: Anxiety and Anger* (1972); and *Loss: Sadness and Depression* (1980). It was a glum set of titles for the introduction, to science, of "attachment," Bowlby's name for the loving magnetism naturally arising between infants and their caregivers.

THEIRS WAS THE SCREAMING ONE, as the orphanage nannies came down the hall with the babies. Eleanor was forty; she felt thirty; she was so ready to be a mother. It was June 2006 and they were in Hubei Province. Eleanor opened her arms to the squalling baby. Despite the wet face and wretched sobs, ten-and-a-half-month-old Lucy was adorable. "Oh my God, she's beautiful," said James, stroking her hair with trembling fingers. All the babies protested, wriggling away from the large white Americans and lunging back toward their caregivers. Lucy looked devastated and confused, glancing back and forth. It was like she was thinking, *I'm being given away again. I thought New Nanny liked me. No one wants me at all.*

Having reached the happiest moment of their married life, James and Eleanor felt chastened by their baby's grief. They carried the distraught child to their hotel room, but couldn't console her. "Lucy, Lucy, Mei Ling," they cooed, shaking little belled toys, offering a bottle. She raised her head on the bed, looked around fearfully—evidently hoping for someone familiar to appear—then put her head down on the bedspread and sobbed. Eleanor and James—in love, in love!—tenderly lifted Mei Ling, swayed with her, and walked the hallway with her. Red patches of anguish appeared on her cheeks. In between bouts of hard crying, she moaned something of a little desperate tune over and over, a fragment of a lullaby from a lost caregiver.

"She's very vocal compared to the others," James murmured as they joined their group for dinner at the hotel restaurant, Lucy wailing in their arms. Other families had started to gel. Their babies smacked their palms upon their plates, splattering the besotted parents. "I think she knows *exactly* what's happening to her and doesn't like it," Eleanor whispered back. "I think she's *extremely* bright."

As days passed, Lucy sank deeper into bereavement. She was listless and had no appetite. The Keiths blamed their inexperience, their ineptitude. But as the baby's mourning stretched into Week Two, they began to wonder: *Is she possibly not okay in some way?*

Lucy wailed throughout the twelve-hour flight across the Pacific. She wept on the domestic flight from California to Alabama. She sobbed as they drove down the state road from the airport. Finally James and Eleanor staggered into their house filthy, deafened, sleep-deprived, jet-lagged, and ecstatic. They had done it! They were a family!

INFANTS COME EQUIPPED with the darling tricks of the baby trade. With big round faces, soft eyes, and moist lips, babies ensnare nearby adults. A child coos or kicks her feet and grown-ups loom above her with warmth and delight. An infant wails in his crib and a parent hurries over and whisks him away for clean clothes and sweet milk. That call-and-response is the foundation of love, of language, and of cause-and-effect thinking. *I do this/ you do that. I scream/you come running! I gurgle/you laugh!*

Yet these innate behaviors can be extinguished. Through no fault of its own, sometimes a baby cannot attract a steady caregiver. Like a little flower unable to summon a bee, the baby wilts. A neglected baby, an institutionalized baby, may smile a tender smile or emit a lonely call of woe with no result. Communication then seems to be a dead end.

An infant who has lost the baby arts of flirtation may finally be dredged from the institutional crib and placed in the arms of eager parents. Now can the sweet call-and-response of love begin?

Not necessarily. Once extinguished, there are instinctive behavior patterns that may not be revived easily. Parental love—a baby's milk and honey—can sour if offered too late. Bowlby noted that a neglected or abandoned child to whom love was belatedly offered sometimes spurned it, appeared scared or confused by it, and could not digest it.

The absence of at least one devoted adult in infancy is a catastrophe for a young child. The missed cues and lost chances can lead to a lifetime of difficulties, which professionals may ultimately diagnose as "attachment issues," "attachment disorder," or (in the extreme) "reactive attachment disorder (RAD)." Diagnoses of post-traumatic stress disorder (PTSD) often accompany the identification of attachment disorder, stemming from the unknown upheavals that left the child, for some period of time, stranded and alone. Lucy would be diagnosed with both.

"We were in love, but she . . . wasn't, not really," Eleanor told me. "I look back at those early weeks and think how naïve we were! Even though we had a really unhappy baby who cried eighty percent of the time, we thought the hardest part of our life was over. We'd read about 'attachment' and felt prepared; but, until we were in it, we had no idea how unprepared we were."

There was no way for them to know how impersonally the baby may have been handled or passed around. "In her first ten months of life, she went through at least three mothers: her birthmother, at least one foster mother (maybe more), and me," Eleanor said. Her physical needs may have been met by a rotation of caregivers, who may or may not have become emotionally invested in her.

When Lucy was eighteen months old, the Keiths moved to rural Iowa, where James joined another automotive firm in engineering and Eleanor moved to the sales division of the cosmetics corporation. Eleanor's elderly mother adored her new granddaughter, and Eleanor's sister embraced her new niece. But then the new grandmother died after a long struggle with cancer and the new auntie died from suicide and Lucy, numbly accepting the news, could add their names to her subconscious list of people who abandoned her. A small child might blame herself for not being lovable enough. No one held on to her. *Why* no one stayed appeared to be the central mystery of Mei Ling/Lucy's life.

At two, she no longer cried all the time, but she didn't fill the house with giggles either. She was a worried child. At three, her anxiety overtaxed her small system. She began jerking out her hair, painfully, clump by clump, until she was nearly bald. The mother fussed over her and arranged cloth headbands, little caps, and scarves. "I don't think she was ashamed about the baldness at first," Eleanor told me. "I think a traumatized child carries a sense of shame inside her. Later she became aware that her hair looked different than other children's."

Her certainty of being rejected again built inside her until she burst out with dismay and outrage. "You're not my mommy!" she shrieked at the mother. "My birthmother is my MOMMY!" Or she'd scream for "Nanny," the name the parents had given the last foster mother. "I need to go live with NANNY!!!!"

To forestall the inevitable rejection by the new parents, Lucy required that everything in life remain *exactly the same every day:* a last-minute change in schedule, a rearrangement of the living room furniture, or an unfamiliar food appearing on her dinner plate seemed to augur the end of her stay with the Keiths. In reaction to a change, in *horror* that her worst fears had come true and that the parents' interest in her had diminished, she shoved away from the table, bellowed and stomped, bit the mother's soft freckled arm, or threw things. In response, the old cat and the elderly dog picked themselves up and exited the room.

When a storm was brewing, Eleanor grimly steered Lucy down the hall and into her pink bedroom—lined with soft toys and pretty dolls and Chinese silks and fans—so that only one room would be tattered in the rampage. In the hall outside her daughter's room, Eleanor leaned against the wall and closed her eyes. She was glad when a rage took place during James's work hours and he could be spared. The child's keening drilled into James's brain like a migraine.

The preschool self-portrait created by Lucy consisted of three wobbly ovals in a row: the first—a tiny circle with four almost-invisible dots for two eyes, a nose, and a mouth—was Lucy; the second, a medium circle with medium features, was Eleanor; the third, a huge wobbly oblong with no face, was her birthmother.

It was as if Lucy were a colander and love the bright water running through it, pooling for a moment before leaking and splashing away.

IN 1989, WITH THE OVERTHROW and public execution of Romanian Communist dictator Nicolae Ceaușescu and his wife, a prison-like system of baby houses and orphanages was exposed to the world. Government policies of compulsory childbearing (no contraception, no abortion, and the decree that women should give birth to no fewer than five children), visited upon a poverty-stricken populace, resulted in a landslide of unwanted births, botched abortions, and abandoned children. After Ceaușescu's overthrow, a hundred thousand children were discovered warehoused around the country in freezing, understaffed orphanages. Starving children wallowed in their own filth in dark rooms; experienced little to no light, sound, or physical touch; and rarely if ever left cribs that doubled as cages.

Visitors observed autistic-like behaviors, self-stimulations, head-banging, self-abuse, self-soiling, and fearfulness about the world outside the crib. Children to whom no one had spoken did not speak; children who had never been rocked rocked themselves. No one could guess their ages, as profoundly deprived teenagers were the size of eight-year-olds, and neglected eight-year-olds were the size of toddlers.

People around the world were stirred to action: relief workers flew in to examine children and deinstitutionalize as many as possible. Hopeful adoptive parents arrived from Western Europe, North America, and Australia. Some visited the bleak institutions to find and adopt children, while others got swept up in an infamous adoption black market operating in the streets, in hotel lobbies, and in rural villages.

The working premise among the new parents of the post-institutionalized children was that *love* would make up for deprivation and abuse. "Love is enough" represented most of what was known about deinstitutionalized children, mostly because the world hadn't seen many like them before.

The post-institutionalized Romanian children took steep turns for the better with their adoptions; most, over time, learned to crawl, to walk, and to speak. They went to school. Some reached miraculous levels of personal and educational achievement.

But some parents discovered that love couldn't reach deeply enough to make all the necessary repairs. Some would write memoirs with titles like *When Love Is Not Enough*. Some of the young survivors continued to be plagued by psychological challenges long after rescue, and many of those issues came to be understood as attachment disorders.

FOUR-YEAR-OLD LUCY WAS TIGHTLY WOUND and apprehensive, a blinking little burglar alarm watching day and night for the warning signs of abandonment. She strived to read the intentions in the parents' faces, in case they were on the brink of abandoning her, and in the faces of acquaintances and strangers, in case they were on the verge of taking her away. "She absolutely needs to know what everyone is thinking at all times," Eleanor said. "When her view of a face is blocked, she panics: her eyes dilate, her face twitches, her breathing gets shallow." Masks terrified Lucy, since they blocked her ability to interpret intentions, so the Keiths gave up on Halloween. Beards upset her, Santa Claus was bearded, and the winter holidays kept Lucy in a constant state of turmoil, of "disregulation." Years of stable family life had failed to chip away at Lucy's uneasiness. Instead of growing more relaxed and confident, she became more frustrated by the transience and unreliability of love. A signal of change—like a new vegetable at lunch—acted on Lucy as a "trauma trigger," the family learned in therapy. "What we perceived in China as awareness was actually hypervigilance," Eleanor said. Hypervigilance is a restless state of high alert maintained by individuals who have learned that the world is treacherous. It can be a symptom of PTSD. Once caught unawares by a terrible event—or subjected to repeated misery—a trauma survivor may brace forever against the return of pain.

The daily mood at the Keiths' was brittle, as happens when there is a family member of any age with a hair-trigger temper. As she grew older, Lucy eavesdropped on the parents' phone calls and refused to "go play" when the parents had guests. On the family computer, she attempted to discover which websites the mother had visited during her absence. Every night's peace was shredded by Lucy's midnight terrors.

Striving for elements of normal life, Eleanor sometimes invited a little girl to their house for a playdate. Though she desperately wanted to have friends, Lucy grew territorial, possessive, dictatorial: "Here, play with this. No, don't touch that." Worries multiplied: *Will the mother prefer that girl to me? The girl is taking attention away from me! That girl wants this mother to love* her! On a playdate at another child's house, Lucy shadowed the child's mother, asking for the escape plans in case of fire or armed home invasion.

"She's adamant that she doesn't want a brother or sister and she can't fathom why any child would want one," Eleanor said. She tried to stop her mother from visiting friends with a new baby and did not want to go. When dragged along, she tried to prevent her mother's carrying gifts into the house. Clearly she suspected: *You want to bring home that baby. You want that baby to love you.*

Lucy was a cute snaggletoothed little girl in pink eyeglasses who mostly "passed" at school as a typical child, unless something horrific happened. When the teacher read *Charlotte's Web* aloud to the class and Charlotte died at the end, Lucy collapsed into grief and despair and didn't care who saw her; if the other children were too blind to perceive Charlotte's death for what it was—an abandonment of Wilbur, who'd loved Charlotte so much!—then they were stupid. But most days she suppressed her anxiety long enough to make it back home from school and through the front door, at which point she'd collapse and rage as if the anxiety were a huge bloodred flag that she'd tucked inside her shirt and could now wave.

Humbled *again,* obliged to seek professional help *again* in their endless search for an everyday family life, the Keiths turned to international adoption doctors, child psychologists, and attachment therapists and tried unconventional treatments, including holistic medicine, chiropractic treatment, equine therapy, neurological reorganization therapy, cranial sacral therapy, and changes in diet. Whatever had happened to Mei Ling had occurred before ten months of age and couldn't be accessed through talk therapy, but diagnoses included developmental trauma disorder (DTD); symptoms of post-traumatic stress disorder, including extremely high anxiety, controlling behaviors, and hypervigilance; trichotillomania (pulling out her hair); obsessive-compulsive behaviors; sensory processing disorder; oppositional behaviors; and night terrors. She was bright—she could count to ten and recite the ABCs before age two. Academics were easy for Lucy, but love was a foreign language. There were happy times together; Eleanor and James treasured moments of silliness and fun with Lucy. "She has so many gifts and blessings inside her!" Eleanor told me. "It's just that the driving emotion for her is extreme fear."

ONE WEEKDAY EVENING IN FEBRUARY, driving eight-year-old Lucy home from her guitar lesson in the neighboring college town but delayed by snowfall and icy streets, Eleanor realized they were approaching one of two trauma triggers: Lucy eating dinner half an hour late or Lucy going to bed half an hour late. Any alteration in a day's pattern would be picked up by Lucy's inner seismograph as a change, and change spelled disaster. Forced to choose between triggers, Eleanor opted to avert hunger by turning into a shopping mall with a dimly lit Chinese buffet Lucy knew and tolerated. Lucy ate dinner on time. But now bedtime would be an hour later. Heading home, in the backseat of the car, Lucy suddenly recalled a neighbor she didn't like. "I hate him! I hate the way he *talks*. Someone needs to cut off his *lips*." The bizarre and cruel thought meant Lucy had been triggered. Eleanor tightened her grip on the wheel and primly said, "Lucy, that's not nice. Don't say anything else." Now she needed to cover the remaining mileage efficiently. If she failed to get Lucy into the house before she exploded, it could take an hour just to get her out of the car. From the moment Lucy calmly said, " . . . cut off his *lips,*" Eleanor figured she had ten to twelve minutes until the meltdown. She squealed into the garage, turned off the car, and hustled Lucy into the house, into and out of the bathroom, and into her bedroom so that when she fell apart, the damage would be contained. In her bedroom, Lucy screamed herself hoarse and trashed the place for an hour.

CHAPTER 6

Casey & Connor

One evening, Scott Millard came across a magazine article about service dogs trained for children on the autism spectrum. Could a dog be Connor's bridge back to life? "What if I call this dog-training agency on Monday, just to get more information?" he asked Deb, and she said, "Yes, do it."

Connor was six years old now, Scott and Deb in their early thirties. Under bristly short hair, Scott's face was all sharp angles and white planes; a distance runner, he had a cautious, darting gaze behind metal eyeglasses. In a short-sleeve button-down white shirt and khaki slacks, he looked edgy rather than geeky; competent; tense. Deb, with the gentlest and most understanding of smiles, still looked about twenty-two, but something steely had come into her soul; she was engaged in a fight to the death to save her son.

The agency had good news and bad news. Yes, autism assistance dogs helped children surmount anxieties and phobias like Connor's. But no, Connor, with his trach and ventilator, was not eligible for a dog. "It's too risky," the staffer said. "Dog hair could get into the air tubes." Over the next few days, Scott made more phone calls to other agencies, without success. In the evenings, he sent emails. After every email, the reply pinged back: "We can't give a service dog to a child with a trach. If something went wrong, you could sue us."

As rejections piled up, Scott considered abandoning the idea. But then he glanced across the room at the frail boy lying propped along sofa pillows, watching *Thomas & Friends* for the thousandth time. The sun was setting beyond a neighborhood stand of Norway pines, pulling all the daylight from the room. "Did you guys get outside today?" he asked Deb.

"We did not."

So Scott pushed on, composing emails to agencies in cities farther and farther west of New Jersey. One evening, he netted a new sort of reply. "Yes, we can help your son."

"I don't know if you saw that our son is ventilator-dependent," Scott dutifully typed back. "He has a trach."

The director, Karen Shirk, replied: "I have a trach."

Trying to read the words aloud to Deb, Scott lost his voice.

SOONER THAN THEY IMAGINED POSSIBLE, they were in posses-
sion of a photo of a one-year-old goldendoodle, a big curly-haired
honey-colored dog with black-button eyes.

Jennifer Varick Lutes had fostered her. "I got her when she
was about a year old," she told me. "I was struck, in the medical
reports, by how fragile Connor was and how much he needed a
dog who was interactive but calm, interactive on a *child's* terms.
A child may be in his own world and you want a dog to draw him
out of it, but not forcefully. Connor needed a best friend. It struck
me how alone in the world he was.

"These aren't really the kinds of things you can train for;
you're looking for the right personality in the dog, one who
enjoys activities that are cuddly more than active. This dog is a
huge cuddler. She's sensitive; she's calm in her reactions. I'd take
her to church with me. I noticed that the smaller the child we
greeted, the gentler she was."

Scott carried the laptop into the den, where Connor reclined
on the sofa. "Connor, look! We're getting a dog! This is your new
dog! Isn't she cute?"

Deb sat down beside Connor to stroke his hair and share the
unusual taste of happy anticipation. "Connor, she needs a name!
What do you want to name her?"

He pulled his eyes from the TV screen for a moment, took
a look, and squeaked, "Ay-ee," which they understood to mean
"Casey," a dog character from *Clifford*.

They realized Connor may have thought they were intro-
ducing him to a new TV character. To bring the idea of Casey
into the physical realm, Scott brought home a stuffed yellow toy
dog for Connor to hold. That weekend the family made an outing to a pet supply store, to buy things for Casey the Real Dog
Who Lives in Ohio Now But Is Coming to Live with Connor
Very Soon. Talking about what to buy had sprinkled their days

with unusually lively conversation. "What color bowl do you think Casey will like? What color leash will you pick for her?" Connor loved gift-giving! He always earnestly created scribble-scrabbled sticker-covered valentines and birthday cards for his cousins, his grandparents, and Christopher. Given the chance to buy presents for a faraway dog named Casey, he tolerated the upheaval in his retirement lifestyle with less protest than usual.

Of course, every excursion with Connor was a space mission. In the parking lot of the pet supply store, the Millards strapped Connor into a jogging stroller retooled by Scott to transport the oxygen equipment, while the usual crowd of trying-not-to-look observers walked past them. The Millard family was always stared at in public, but rarely spoken to, other than the occasional "What *happened* to him?" or "What's *wrong* with him?"—usually from a child, right in front of Connor.

If bystanders had ever broken through the imaginary wall of spectacle to make contact, they might have been told: *We didn't know a thing more about all this when we started than you do now. You just do what you have to do to keep your child alive.* Or: *We love him. This is our son, our only child. We do this out of love for him.*

WHEN SCOTT FIRST SAW CASEY the Real Dog at 4 Paws—taller and longer than he'd imagined, and so covered in blond fluff that she looked like a giant Muppet—he thought, *I have made the worst mistake of my life.* Heeling to a young trainer, Casey sashayed into the room, her furry feet swooshing along the linoleum like big square house slippers. Scott stiffened. He didn't dare glance at Deb, in case she shuddered, too. Instead of lessening their immense daily burden, he realized, they'd just multiplied it. He hadn't realized what a wave of hope he'd been riding until he felt it collapse beneath him at the approach of the dog.

The Millards had caravanned to Xenia with Deb's parents, Dick and Dorrie Tiel. Now Connor rested at the motel with his grandparents, while Scott stood in the training circle trying not to flinch at the approach of their frisky new family member. Unlike the other failed experiments, this brilliant idea (*his* brilliant idea, to be honest), the wiry jubilant animal lunging toward them across the training circle, could not easily be stashed in basement storage. *Can we just acknowledge our mistake and leave now, without the dog?* Scott wondered. *Lord, now lettest thou thy servant depart in peace.* They could swing by the hotel and make it to the highway within the hour. But if they bailed, what would they tell the friends, congregants, and strangers who'd opened their checkbooks and bank accounts, who'd handed them cash for gas, meals, and a hotel, without expressing what they must have been thinking: *A dog? For _Connor Millard_? Why not a gerbil? Why not a fish?*

Like trapped men everywhere, Scott wiped his palms downward along his pants legs, lowered his head, and flicked his eyes toward the exit.

Casey flounced up to them, eager and vivacious, pulsing with energy. Wagging all over, tongue splashing out, she banged against their legs. In shock, they patted the springy thickly padded hairdo. Scott took the leash and offered, "Casey, sit?" and she didn't sit but continued to wiggle about excitedly, panting and whacking their knees with her tail.

Scott thought again: *Oh my God, what have we done?*

Deb thought: *This dog is going to kill Connor.*

AT FOUR O'CLOCK THAT AFTERNOON, they were relieved when trainers collected the dogs for the night. To the Millards' surprise, other parents knelt and murmured endearments to their dogs, ruffling the fur around their necks, using their cell phones to take the last photos of the day; and some of the children—children with Down syndrome, with autism, with dwarfism—threw their arms around their dogs' necks and hugged them goodbye. Even in this group of special needs families, the Millards were in a class by themselves, only Connor too fragile and anxious to come to class, only Connor at risk of having his narrow life worsened by the addition of a dog. The Millards practically flung the end of Casey's lead to a trainer, headed for the parking lot, and didn't look back.

"How did it go?" asked the Tiels when Deb and Scott returned to the motel room. Deb looked frazzled, unfocused. Scott toppled full-length facedown onto the nearest double bed. "Where doggie?" rasped Connor.

"She wasn't quite ready yet, but she'll come meet you tomorrow," said Deb.

"Okay," Connor said cheerfully, content to snuggle with Casey-the-stuffed-dog and in no hurry to get close to a real animal.

From the pillow into which Scott's face had disappeared, there came a low moan.

"SHE WON'T SIT," THEY TOLD Jeremy Dulebohn the next morning. Deb, smiling anxiously, trying not to sound critical, couldn't help but express their greatest concern as Casey romped around their legs: "She's kind of like a big puppy, isn't she?" Scott emitted a dry laugh. Deb felt unready and scared to return home with the dog.

Jeremy was used to this. He often said he wished he and the families had six weeks together instead of two, so that the parents—all novice dog-handlers—could approach the level of knowledge and training the 4 Paws dogs had reached. The dogs responded to the precise commands and crisp gestures of the 4 Paws trainers, but their new owners were all over the place, sloppily throwing around extraneous hand movements, tossing in extra words, phrasing commands as questions. Jeremy said to Casey: "Sit." Casey sat.

Jeremy now escorted Casey into the center of the training circle. Suddenly they were like Olympic skaters in a pairs competition, intricately attuned to one another's pure and graceful movements. A transformed animal, calm and intelligent, Casey glided at Jeremy's side off-leash, heeling perfectly, turning when he turned, dropping to the floor when he told her "Down," finding and sitting on her mat when he said "Place," rejoining him when summoned, swinging her body around to resume the "Heel" position.

Deb thought: *Wow.*

Scott thought: *This might not have been the worst mistake of my life.*

Their performance completed, Jeremy and Casey cruised back to the outer circle. "She's trained," Jeremy said with a kind smile, handing over the lead. "You're not trained—yet—but you will be. That's what we're doing here. We're not teaching her. We're teaching you."

Now Deb and Scott took turns playing the role of Connor. Scott, on the floor, stiffened, trembled, gasped for breath, and pretended to silently rage. Deb said, "Touch." Casey gently drew near to Scott and put her paw on his leg.

Then Deb played Connor going into a pouty lockdown of silent refusal. Scott said, "Lap," and Casey insinuated her big head into Deb's lap. The hours flew! They felt charmed by the big ridiculous-looking dog, who seemed eager to comprehend and to please them. She'd start lowering her haunches in advance, watching their faces, in case they were about to say "Sit," but if their lips formed the word "Heel" instead, she hastened to whirl her curly self into position, then looked up at them to see if they were pleased.

"This is a dog who responds to praise more than to treats," Casey's trainer, Jennifer Varick, told the Millards, and they felt it. When they said, in warm tones, "Good girl!" Casey did a sideways soft-shoe in evident pleasure. She had an open smiling face. When she lay at their feet in the "Down" position, on her mat, Deb and Scott casually leaned forward in their canvas folding chairs and twisted her ringlets. Rumbles of contentment emanated from the dog. She was almost purring. Occasionally she turned around and gazed up at them, as if wondering, *You?*

They could fall in love with this dog! But the plan was for Casey to bond with *Connor,* not with them. "Don't let your dogs get overly attached to you, parents," Jeremy reminded every class. "Give them affection, but keep leading them back to your child." Deb secretly wished to be Casey's favorite. So did Scott. A few extra treats smuggled her way might do the trick. But they stopped themselves. Connor needed her more.

THEY ENTERED THE MOTEL ROOM cautiously, calling in soft, excited voices, "Connor! Casey's here!" Connor, who hadn't been feeling well, awoke groggily from a nap in his grandparents' room. Deb told Casey to hop onto the double bed and assume the "Down" position, while Scott stepped into the adjoining room.

Connor's face was one big worried smile as his father carried him across the threshold; the closer they got to the unkempt mischievous-looking animal on the bed, the tighter the stranglehold he kept on his father's neck. If he could have crawled up and perched on top of Scott's head, he'd have done it. "Down, down . . . ," the Millards kept reminding Casey.

Connor allowed himself to be lowered inch by inch onto the foot of the bed; very slowly he slid out of his father's arms. Boy and dog stared at each other, both with mouths partly open. Casey whipped her tail rapidly but otherwise stayed low and still. Scott and Deb were afraid to move, fearing that an accidental hand gesture would inspire Casey to leap onto Connor and lick him. That would be an epic mistake. They shushed the grandparents, who came bustling into the room to watch. It was a silent tableau.

Then Casey scooted forward a millimeter. Her haunches didn't budge; she just extended her neck. Connor stuck a small shaky hand into the air. Casey stretched forward until the little hand could sense the haze of her brass-colored fur. Feeling the tickle on his palm, Connor mechanically patted, as he'd practiced patting Casey-the-stuffed-animal. A low gurgling sigh came from Casey's throat. Connor laughed. Deb and Scott, tremendously relieved, burst out laughing, and Deb's parents bent over laughing, too.

Casey's gentle temperament and excellent training had carried the family over the hurdle of first encounter! Her temperament and training and . . . was there anything else? Did she intuit that this was a child? That he was frightened, and breakable? Maybe even that he was *her* child, the one she'd been in training for? Scott knew this was what Karen Shirk and Jeremy Dulebohn labeled the Lassie Myth—the faith that their new dog would rescue their child and fix everything. And yet, Scott did feel they were safe—that Connor was safe—with Casey, that Casey was stage-managing this interaction.

In any case, they didn't need Casey to fix *everything*. They just longed for the dog to give Connor a taste of childhood, a bit of fun and friendship.

Connor grew tired and asked to watch TV. "Can Casey watch with you?" asked Deb.

"Yeah."

They fluffed up the pillows, turned on the TV, and lost Connor into the transfixing animated landscape. While Connor focused on cartoon animals, Casey focused on Connor. She inched closer and closer, almost imperceptibly, until her nose came to rest a quarter-inch from Connor's hand. Her warm exhalations puffed against his fingers. The parents turned to other activities but, when they glanced again at their son, they saw that Connor's hand rested on Casey's head.

That evening, Deb asked Connor to pour food into Casey's dish. "Connor, she's so hungry!" She helped him descend from the bed and wheeled the metal cart beside him as he toddled into the bathroom. He tumbled kibble from a cup into a dog dish and stared, amazed, when Casey bent to eat it. "Good job, Connor!" they said. "Let's give Casey water now." Connor slopped water from a plastic cup into her water bowl on the tile floor and Casey lapped it up.

They slung across Connor's shoulder a little 4 Paws pouch in which treats were held. They'd practiced various commands with Connor for weeks at home. "Connor, can you tell Casey to sit and give her a treat?"

" 'it, Ay-ee," he whispered. She didn't hear him.

"Louder, Connor; try again."

" 'it, Ay-ee."

Still too soft.

He drew in his deepest breath and gave his loudest squeak: " 'it, Ay-ee!"

She sat!

He reached into the pouch, selected a tidbit, and shyly held it out. Casey slobbered it from his fingertips. The Millards froze as Connor studied the wetness on his fingertips; at home, smaller indignities than this could trigger an hour-long freak-out. Now he held up his fingers for his mother to see.

"I know!" she laughed. "Casey kissed your fingers when you gave her the treat."

He nodded soberly.

" 'own, Ay-ee," Connor said. Casey lay down. Connor chortled and fed her a treat.

" 'eak, Ay-ee."

"She can't hear you, Connor."

" 'EAK, AY-EE!" Casey sat up, swallowed, lifted her face, and yipped. A flinty gleam of authority entered Connor's mild eyes.

Unlike most boys his age—who rode bikes, joined the Cub Scouts, went to birthday parties, cannonballed into swimming pools, played baseball or soccer, and started first grade—six-year-old Connor had no arenas in which to feel competent, important, or powerful. He wasn't even toilet trained. Suddenly a magical downy creature—who stood eye-to-eye with him in height—showed up, listened closely to him, seemed to like him, and obeyed.

Watching the two interact, the Millards felt like Casey *opted* to obey Connor. She was a strong young animal, just fifteen months old, and she outweighed Connor by fifteen pounds. With a swipe of her tail or paw, she could have knocked him down and bolted from the room; instead, with bended head, she complied with the wishes of the bossy barefoot pipsqueak in Thomas the Tank Engine pajamas. It felt to Deb and Scott like they were in cahoots with Casey to make Connor feel powerful. "Thank you, Casey," Deb murmured to her at bed-time, stroking her broad head where she lay on the bed beside

Connor. Without getting up, she replied with a few thumps of the tail.

THEY HADN'T THOUGHT ABOUT HOW LONELY Connor must have been, in his self-imposed seclusion. Now, at home, he sought out Casey a dozen times a day. When Casey sprawled on the floor, unwinding after her early-morning jog with Scott, Connor climbed down from the sofa, hobbled over, and threw himself on top of her. Or, resting on the couch, he squeaked her name and she careened into the room, nuzzled into him, jumped up, hogged the sofa, and made him laugh. When, in a sour mood, Connor angrily thrashed about and threw things, Casey cheerfully wandered into the kitchen for a pat and possible snack from Deb, or she flopped onto her bed in a corner of the living room to catch some shut-eye. When Connor felt better, he invited Casey to play trains with him; he picked out which engine and cars she could use (not his best ones) and put them in her paws while he set up the track. Alas, in train races, *her* train always lost. "Too bad, Ay-ee," Connor snickered.

Late one night, Casey came into the master bedroom and barked. Deb and Scott awoke and sprinted to Connor's bedroom. He was fine, all systems chiming and whirring. "No, Casey," they said, and got back in bed. As they closed their eyes, she barked again. "I'll go," said Scott, and staggered back to Connor's room. But Connor was breathing, the tubes were not tangled, the engines were purring. The house doors were locked and there was no odor of smoke or gas. "No, Casey. Go to bed," he begged. As he got under the covers, she barked again. Deb sat up to think, looked around the room, and noticed a problem. It was subtle, but crucial: the green light on the baby-monitor receiver on her bedside table was off.

Whenever Connor wasn't next to them, the Millards listened to him through a sound monitor. That afternoon, while Connor napped, Deb had taken the receiver with her into the garden. At bedtime, she'd placed it back on her bedside table, but evidently hadn't plugged it into a wall socket. In the night, the battery ran down and the receiver died.

Deb plugged it in. The small green light beamed. Casey did an about-face and left the room. WHAT? Deb and Scott looked at each other skeptically. Had Casey really just alerted them that the sound monitor was off? Out of dozens of lights blinking in this twenty-first-century American home—computers, cable box, coffeemaker, microwave, digital clock radios, cell phones, and the medical electronics surrounding Connor—she'd alerted to *this* one, not even in Connor's room? Was it possible there was a supersonic sound given off by the receiver and Casey noticed its sudden absence and felt compelled to tell them about it? A dog understanding electricity, battery power, radio waves, and/or the significance of parents monitoring a medically fragile child remotely? Ridiculous! No one would even believe this. But it happened.

"CASEY WANTS TO GO OUTSIDE AND PLAY," Deb said, hoping to break up a long indoor afternoon of TV-watching, meds, and napping with a trip trough the kitchen door to the back deck.

Despite his terror of trees, wind, clouds, puddles, and planes, Connor said, "Okay."

"Do you need your sunglasses and headphones and . . . ?"

"No."

Connor stood beside his equipment cart—bareheaded in the sun and wind, exposed to the elements!—and threw the tennis ball as far as he could. It fell in front of him and dribbled off the deck onto the yard. Casey made do with the thirty-six-inch throw; she raced around and circled the ball, nipped at it, picked it up, bounded around the yard with it, dropping and retrieving it several times, then trotted up the steps to drop it back at Connor's feet. Connor threw back his head, squinted his eyes shut, and laughed long and hard. "T'row again!" yelled Connor, and heaved the tennis ball a good thirty-seven inches into the grass.

Casey flew off the back deck in pursuit of a squirrel, lost it chattering over the fence, and returned huffing to Connor's side. The vicarious pleasure of the chase made Connor kick his little feet in excitement. After an hour he came back into the house ready for lunch and a nap, having truly expended some energy watching and rooting for Casey.

"Can we take Casey on a walk?" asked Deb one day, and Connor—forgetting the risk of encountering birds, trucks, squirrels, lawn sprinklers, strangers, and clouds—said, "Yeah." Fitted with a harness and rubber handle, Casey was a prop for Connor, steadying his off-kilter walk. Holding on to the harness, Connor wobbled down the driveway and up the sidewalk several hundred yards before growing fatigued. Casey ambled smoothly beside him, supportive in every way.

"Going to the doctor today, Casey's coming," Deb would call, and Connor would forgo the tantrum. Or, if he started to rage, Deb would say, "Casey: Touch." Casey would lay her big paw on Connor's leg, reminding him to calm down.

One day that fall, for the first time ever—rather than being carried in by a parent or pushed in the modified stroller—Connor walked into a doctor's office on his own two feet. There—another first—people talked to him! To Connor! "Is that your dog?" a nice woman in the waiting room inquired.

Flabbergasted, Connor looked to his mother to learn how to handle this unprecedented circumstance.

"You can tell her that's your dog."

"Yeah," said Connor.

"What's her name?"

Again he looked to his mother. "You can say her name is Casey."

"Ay-ee," whispered Connor.

"Oh, 'Casey,' what a nice name!" the woman said. "She looks like a very nice dog."

And Connor, abashed and pleased and proud, looked down, smiling. Alone amid his tubes and pipes, he was invisible. With a big ruddy dog at his side, he was somebody; he was a *person*.

For a year, a speech therapist had come to the house to try to inspire Connor to speak more than single words at a time. Deb listened one day from the kitchen as the poor tutor, at the dining room table, struggled to extract multiple-word sentences from Connor. Connor was bored, or tired, or unhappy, or rebellious; Connor had nothing to say today. As it neared the top of the hour, Deb spared the poor woman continued frustration. "Thanks for trying," she said kindly, and walked her to the front door. The two women stood in the foyer momentarily, sharing some hopes and strategies for the future, when Deb suddenly held her fingers to her lips, shushing the guest. "Listen!"

From the den, they heard . . . like the busy whistling of a bird on a branch in springtime . . . a high-pitched breathy monologue, a squeaky soliloquy . . . Connor was talking, Connor was babbling, to Casey! They sat side by side on the den sofa—Casey, resting her head on her front paws, gazed off into the middle distance, while Connor looked down upon her from above and held forth. The mumble of little whispery syllables included, frequently, "Ay-ee," followed by a deep breath, and then another arpeggio of nasally notes. The mother and the speech therapist couldn't make out the subject, but they perceived emotion, syntax, punctuation, narrative arc, rising tension, and perhaps even denouement. Since Casey's arrival, Connor had worked hard to speak loudly and clearly enough for his commands to be understood; now he seemed to have grasped the essence of speech as a medium for relaying one's innermost thoughts and feelings to one's closest friend.

A thunderstorm rumbled low over the house one night, pounding on the roof and banging at the windows. The bedrooms lit up with lightning. A high, frightened wail from Connor sent Deb and Scott hotfooting it into his bedroom. There they found the boy sitting up in bed with his arms around his equally terrified dog. Both child and dog looked up at the adults with eyes wide with fright. Casey was shaking. Connor had screamed for his parents on behalf of them both. Deb and Scott burst out laughing. They settled onto the bed, cradling Connor and Casey between them, snapping the bedspread over the scared duo like a tent. All four fell asleep crammed together there till dawn. Casey had already become Connor's best friend and playmate, his protector, and his coach; at this moment he was effectively Connor's little brother. Every worldly relationship denied Connor by illness was somehow offered by this soft goofy dog.

One weekend, Deb and Scott pushed Connor in his stroller, with Casey trotting alongside, to the nearby school playground. Connor got out, grasped Casey's harness, and (while his dad wheeled his breathing equipment cart beside him) tottered a few steps onto the blacktop, toward the slide. Kids at play noticed the interesting group, including the fact that the big dog wore a red 4 Paws for Ability vest. They ran over in friendly curiosity, just as Karen Shirk had said they might. "Is that your dog?" "Is it a boy or a girl?" "Can I pet her?" "Can she do tricks?" "What's her name?"

"Yeah!" said Connor. "Ay-ee . . . 'it, Ay-ee."

Casey sat.

"What else does she do?" "Can she shake hands?"

" 'ake, Ay-ee."

The kids laughed and shook hands with Casey.

"Casey can give you a high five," offered Scott.

"Fi', Ay-ee," said Connor.

The children took turns slapping the plump rough paw-pads in midair. "I like your dog!" "Bye!" "Thanks!" "Bye, Casey!"

Connor mouthed the word, "Bye."

At bedtime that night, Deb asked her son, "Did you have fun today?" (*"Did you have fun today?" What a thing to ask Connor!*)

"Yeah!" said Connor. "Ay-ee."

CHAPTER 7

Karen & Ben

One afternoon in 1989, Karen Shirk, a twenty-six-year-old social work student at Wright State University in Dayton, Ohio, collapsed onto the pavement, spilling her books and purse, scraping her elbows, and knocking her chin as she went down. She hadn't tripped; she had simply flopped down like a popped balloon. She lay stunned, facedown, breathless. People jostled around, yelling as if from far away: "Are you okay?" and, to each other, "Is she a student?" In blue jeans, gray sweatshirt, and white Keds, her short straight brown hair tossed over her face, Karen was trying not to suffocate. Her eyes were wide open but she couldn't capture the breath to say anything. No one recognized her. Someone ran to a building in search of a pay phone, to call an ambulance.

Karen was earning a bachelor's degree in social work, while working full-time at a day care for developmentally delayed adults. She'd experienced greater than usual fatigue in the weeks leading up to her collapse, but chalked it up to her long hours and overfilled days.

To others, she looked the part of the future social worker: short, peppy, and round-cheeked, with a bustling short-legged stride and a bowl haircut. She played the part, too: cheery and affirmative, offering an easy bark of laughter in support of a professor's or classmate's attempt at humor. She drove a dinged-up Chevy Nova and shared a rental house a few miles from campus. Occasionally on the receiving end of a general invitation as an evening class broke up, she might gamely traipse along to a neighborhood tavern. She didn't say much, so it was easy for others to overlook the quiet rage of intelligence or the occasional double take of skepticism. At a young age, and without reading glasses, she'd somehow already perfected the down-tilted, over-the-top-of-the-spectacles look of *Oh, really?* But no one was paying attention.

Quietly sipping from a mug of draft beer at the damp table, she took pleasure in enacting a classic college scene but knew she wasn't indispensable to it. Born to Captain and Mrs. David Shirk on a U.S. Army base in Frankfurt, Germany, in 1962, the first of four children, Karen grew up, she says, "in Germany, Virginia, New York, Arizona, North Carolina, New York again, Kansas, Germany again, Indiana, New Hampshire, Ohio, and I'm not sure that's a complete list. Fort Huachuca, Fort Bragg, Fort Riley. I never saw any of them as 'home.' " When she came in after school to find the cardboard boxes set up in the living room again and a suitcase open on her bed, there was no point in mentioning that a Girl Scout Brownie troop was forming and she'd brought home a flyer.

The one constant in her childhood was Moonko, a black German shepherd puppy her father brought home when Karen was a baby. "*Moonko* felt like home," she told me. But as she and Moonko reached their eighth birthdays, the family's itinerant lifestyle took its toll on the dog. When Karen's father left for a tour of duty in Vietnam, Moonko snapped. He tried to block family members from leaving the house and warned strangers not to enter. Karen's mother felt she couldn't cope and gave Moonko to a veterinarian who'd admired him. Karen's father returned from Vietnam and the family packed for the next move, but there was no mention of retrieving Moonko. He was swept away from Karen like everything else.

By ten or eleven years old, she'd perfected the neutral, mildly amused outsider stance of the "army brat." Oddly, it became her attitude even within her own family. In her twenties, she was sometimes curious about the close bonds people seemed to share with their families and their close friends—she didn't really feel like that. She could take social life or leave it. There were a few students on campus whom she saw regularly, but she didn't think of them as friends exactly. She thought of them as "weirdos," like herself, and she was fine with it.

She was also compassionate, though she hadn't had much chance to express that side of her nature. The barks of laughter she emitted over human foolishness, narcissim, and arrogance (which, in the internet age, she would display online with comments like "BWAHAHAHA") never left her, but there were hidden wells of kindness. She had been an outsider, a social isolate, most of her life. She never expected anyone to be that interested in her or even that nice to her. She was years away from discovering that, were she to find herself on top of the world, she would be the type who reaches back to folks in worse straits, in greater pain, than herself, rather than swing along solo from success to success. It was too soon to know this.

BY THE TIME THE EMTS ARRIVED, Karen Shirk was in respiratory arrest. That night, tucked into a hospital bed in intensive care, fitted with an oxygen mask, in her mind she kept falling, as if she'd not come to rest on the sidewalk but had plunged through it. Under a thin sheet and blanket in the freezing ward, she waited for the nurses and doctors to quick-start her muscles and lungs again. But nothing worked. Her parents and siblings turned up at her bedside, and left, and came back, and left. Tests were run; theories and diagnoses were floated. Karen's learned passivity toward sudden life changes served her well: she knew she had to ride it out, no matter how difficult; she knew it was futile to protest. She didn't have the air to voice a complaint anyway. She pulled deeply into herself, keeping a small flame of Karen Shirk alive while appearing alarmingly unresponsive to everyone else. A couple of the "weirdos" visited, then forgot to come again.

Hospitalized for months, all systems failing, Karen finally received the grim diagnosis of myasthenia gravis (MG), a rare neuromuscular disease known for sudden onset in the twenties. By twenty-seven, she was a respirator-dependent long-term disability-housing resident with nursing services and no hint that life would improve.

NEARLY ALWAYS ALONE on a hospital bed in front of a television that ran nonstop, Karen seemed unaware of the hour, day, or month. "Why don't you get a service dog?" asked Alice, Karen's personal care attendant, six years into Karen's residency. A middle-aged woman with strong freckled arms, a shiny blond hairdo of tight ringlets, and a no-nonsense, upbeat attitude, Alice smelled of cigarette smoke and spearmint gum. She came from generations of Ohio farming people and knew the length of a day's work. Appraising Karen Shirk, she wondered if a service dog could jump-start the life of this nearly forgotten, bloated, weak, and silent young woman who'd washed up here, far from her previous life or attachments. A service dog could be trained to offer mobility assistance and to do small tasks, Alice told Karen, like open and close her bedroom door, turn her light switch on and off, and bring her things. Karen thought Alice was bossy. She asked to be left alone. But Alice, impatient with every kind of indolence, even medically induced lassitude, had made up her mind that Karen needed a friend.

"INTERSPECIES FRIENDSHIP" IS A POPULAR SUBJECT these days. Books, websites, calendars, and movies abound with "remarkable" interspecies friendships and "unlikely" interspecies friendships. In the children's book *Owen & Mzee: The True Story of a Remarkable Friendship,* one learns of Owen, an orphaned six-hundred-pound baby hippopotamus in Kenya, who insists that Mzee, a 160-year-old male Aldabra giant tortoise, is his mother, and Mzee appears to agree. In Jennifer Holland's bestseller *Unlikely Friendships: 47 Remarkable Stories from the Animal Kingdom,* we meet the Macaque and the Kitten, the Papillon and the Squirrel, and the White Rhino and the Billy Goat. Photos orbit cyberspace in which a baboon grooms a mountain goat, five squirrel monkeys sit on a capybara, a crow takes care of a kitten, and a sheep naps with an elephant.

In a ludicrous photo now making the rounds, a raccoon slides through a Florida lagoon on the back of an alligator. No one argues "friendship" for this one, though. Someone posted, astutely I thought: "Because Florida."

NBC News reported the story of a four-foot-long rat snake, Aochan, living in the reptile house at Tokyo's Mutsugoro Okoku zoo, who began to boycott his dinner. He did not like frozen mice as an entrée. Concerned about him, his keepers one day sacrificed a young dwarf hamster, three and a half inches tall, whom they named, with dark humor, "Gohan," the name of a tasty rice dish. The snake swirled around the victim, flicked its tongue, tightened the noose . . . and then closed his eyes for a nap, while Gohan made a nest for himself within the snake's smooth coils and caught a bit of shut-eye himself. Aochan has never eaten Gohan. The snake now accepts frozen rodents for dinner. "I've never seen anything like it," said zookeeper Kazuya Yamamoto. If Aochan had shared his enclosure with another snake, perhaps the pair of them would have feasted on tiny Gohan. But, in the absence of snake companionship, it appears that this snake needed someone to digest less than he needed someone with whom to pass the long winter nights.

I don't find these stories all that remarkable or unlikely. I find them profound and poignant because of how universal and *likely* they are. They reveal how deeply wired all living things are to bond with a companion.

Every creature is born with a set of instinctive movements and features and skills—winks, kisses, smiles, nibbles, pheromones, sultry glances, tail wags, slime trails, purrs, whispers, buzzes, hoots, feather displays, antler-shaking, lights that glow in the dark, whistles, swollen chartreuse butts, flops of the flipper, bedroom eyes, and gift-giving and nest-building—to attract a mate and to be welcomed as a friend and comrade into the pack, pride, herd, litter, covey, bevy, gaggle, flock, school, swarm, or tribe.

In great flocks, migrating birds cross the sky, wheeling in unison; more delicate than origami, hundreds of trembling butterflies touch down on a field of flowers; with a flick of a tail, a fish alters the route of hundreds of his gleaming brethren. The latest research on cave slime marvels at how well the microorganisms—looking like the dirty bubbles of sea foam left behind by retreating ocean waves—work together. They join arms (so to speak) and go exploring. They solve mazes presented to them by researchers! In the wild, those on top of the gelatinous heap sacrifice their lives for their brethren on the bottom.

No animal evolved a repertoire for permanent solitude. No species spent millions of years figuring out what to do when home alone on a Saturday night. Such an animal would never have reproduced. Our stratagems are all about looking cute, joining in, nurturing and being nurtured. Is it really so remarkable that if a member of our own species isn't available, we'll bond with a member of another? Not really. We tend to follow the song's advice: "Love the one you're with!" We will run through our gamut of charms for another creature, regardless of its kingdom, phylum, class, order, family, genus, or species, and—if that creature is willing—we can be happy together, because it turns out that *attachment* is the thing.

If a human being's desire for closeness with another human being is stymied—perhaps by physical, emotional, or cognitive disability—a friend may be found within a population other than *Homo sapiens*. Even if the new friend arrives on four legs, sheds a lot, drools, and has really bad breath, the result can still be an explosion of *love,* mind-expanding, healing, and joyful.

"HOW COULD I TAKE CARE OF A DOG?" Karen rasped through the permanent tube in her throat. "I can't even take care of myself." She was thirty-two years old and had spent 313 days of the previous year either in the hospital or in assisted living.

"You could take care of a dog," Alice said crisply, turning off the TV and hoisting the window blinds. Karen groaned in protest. Alice was no scientist and thus held no scholarly opinion about dogs as a man-made species who lacked a biological niche.

She had a working person's healthy respect for them. Operating on the border between optimism and despair, Alice had seen dogs pull more than their weight in dragging their humans across the boundary line toward greater health, love, and life.

With nothing else to do, with her days a monotonous series of TV shows and tasteless meals on trays, aware that she was growing feebler by the week, Karen one day struggled to push herself up a bit higher on the pillows. After half an hour of work, sweating and out of breath, she achieved something approximating a sitting position. Weeks earlier, Alice had phoned the American Animal Hospital Association for a list of service dog providers. When the list arrived by mail, she had wheeled an over-bed table in front of Karen and placed a telephone, writing paper, and a pen on it. Karen had ignored it all—tray, telephone, pen, and paper. But now she reached feebly for the tray, studied the list of agencies, mailing addresses, and phone numbers, and made her first call. It was the most productive she had been in six years.

Daily, over the phone, she croaked and whispered her situation, or expressed it in writing. To all, she mentioned her lifelong love of dogs.

In time, by phone and by letter, the agencies rejected her, some within minutes, by phone, while others took a few hours, days, or weeks to call back or write back their denials. Their answers were the same: "We do not place our dogs with respirator-dependent individuals who will never lead a productive life." Or: "Others with greater abilities will make better use of our dogs in their reintegration into the community."

Karen slid down on the bed again, eyes back on the ceiling. In the office down the hall, Alice got on the phone, too, looking for more agencies to contact.

Then an agency said yes! They placed Karen on the bottom

of their waiting list. She began to feel something like hope. She didn't object when Alice twisted open the blinds and allowed sunlight to stripe the room in yellow. Her food regained a touch of flavor. A year passed. And another half-year. Oh no, you're still on the list, the agency assured her. Finally, eighteen months after her acceptance, a trainer arrived to assess Karen's needs personally and to prepare her for the placement of a golden retriever. She and Alice answered all the visitor's questions with great earnestness and commitment. The trainer left, promising to return with a dog in two weeks.

Two weeks later, instead of a dog, a letter arrived. In shock, Karen read it aloud to Alice: "Our guidelines prohibit the placement of service animals with people on ventilators." Evidently they'd misread her application a year and a half earlier, in which ventilator-dependency had been clearly stated.

Karen was finished. *That's it. I don't care if I live or die. This is nothing but a long slow decline toward death.* She began setting aside painkillers, stockpiling morphine.

"KAREN," SAID ALICE ONE DAY the following year, 1996, brooking no debate, fed up with the service dog agencies and fed up, apparently, with Karen, too. "Get your own damn dog."

"I don't know how to train a dog," croaked the patient, flat on her back, years wasted, secretly counting pills against the day she'd have enough to make an end of it.

"So hire a trainer."

"How would I even get a dog? Are you going to find me a dog?"

A few days later, Alice, who had become her best friend, her only friend, got bossy again. "Karen, get up. Let's go see some puppies."

"What puppies?" said Karen.

"I don't know! Look, the newspaper ads have lots of litters. Let's go look at them. We'll just look—you don't have to get one. Ha! Maybe I'll take one!"

Feebly, with the stronger, older woman's help, Karen got dressed in bed and crumpled forward into a wheelchair. In blinding sunshine, Alice wheeled her up a ramp and into a van, secured her, and drove to the first house. There were puppies in a pen in the garage. Alice knelt on the spread-out newspaper to play with them, and reached up high to place a squirming brown-and-white one in Karen's lap, but nothing stirred in Karen. She was in a trance of decline and doom. Years from now, she would call this time in her life "the Days of Death." "I'm ready to go," she told Alice, and called "thank you" in a high monotone to the breeder as Alice pushed her back to the van.

"Didn't you say your favorite kind of dog was a black German shepherd?" Alice tried another day. "You're going to go look at puppies."

Again came the tremendous undertaking of getting Karen dressed and out of her bed. Alice couldn't take the time off work that day and had asked another personal care attendant to drive Karen to a breeder's house. In the backyard, Karen sat in her wheelchair facing an outdoor pen crawling with black German shepherd puppies. They tumbled forward. "The three little girls all ran to the front, jumping up, like *Pick me! Pick me!*" Karen remembers. "The little boy just sat there with an expression like, *I'm the one you need.* So I said, 'I'll take that one.' "

She named him Ben.

And that was the end of her hours, days, months, and years alone in bed. The puppy had to be taken outside, cleaned up after, fed, brushed, bathed, played with, and walked. He liked to go to parks and run. He liked to pounce on a scrap of paper or a dandelion and play at breaking its neck. After an outing, by the time Karen had heaved herself out of the motorized wheelchair and back into bed, breathing hard through her tube, Ben frisked about her room, ready to go out again.

Oh my God, she thought. *This is too much. What was I thinking?*

She asked Alice, "Could you take Ben out?"

"Ha! Too busy for that, my friend," said Alice.

Karen gasped for breath and strained long-unused muscles. She rolled and slid off the side of the bed, crawled and climbed back into the wheelchair, connected Ben's leash, and motored feebly out the front door while he pranced alongside. "I didn't leap back into life with Ben so much as inch back into it," she told me.

Loneliness was gone, boredom was gone, self-pity was off the table. The TV was off and the windows were wide open. In front of the apartment building, when the bit of black fluff cavorted in the grass, strangers approached the large, pale, tubed-up woman in the wheelchair with friendly conversation, as no one ever had when she was alone. It was a lesson Karen wouldn't forget.

She registered for a puppy obedience class and managed to get herself and Ben out the door, into a wheelchair-accessible van, and across town. By the time year-old Ben graduated from puppy classes, he was a gorgeous animal with a shiny coal-black pelt, orange-flecked brown eyes, and a feathery tail; and Karen had become the owner, with a grant from the Bureau of Vocational Rehabilitation, of her own van with a wheelchair lift. New medications allowed her to come off the ventilator during the day. Now she wanted service dog training for Ben. She took him to the National K-9 School for Dog Trainers in Columbus, where they met a twenty-three-year-old trainer from Wapakoneta, Jeremy Dulebohn.

Since a morning in Jeremy's third-grade classroom, when a police officer from a K-9 unit visited and demonstrated a German shepherd dog's ability to sniff out contraband, Jeremy had planned to become a police officer. He'd never swerved from that path until, upon graduating from high school, he learned that not all officers worked with dogs. Rather than apply to the police academy, he had applied to the National K-9 School for Dog Trainers. Now he taught Ben the basics of mobility assistance: to open doors and drawers by pulling on ropes; to fetch Karen's wallet from her lap, stand up at a store counter, drop the wallet, and then return it to Karen with the change inserted by the retailer; to brace her for balance as she moved from bed to wheelchair and back; and to remove her shoes, socks, and jeans at bedtime.

Back at her room in the assisted living facility, when Karen asked for water, Ben opened the refrigerator and pulled a cold bottle out with his teeth. When she asked for laundry, he pulled the clean clothes out of the dryer into a basket and dragged the basket across the floor to her. When the phone rang, he waited for her command, because sometimes she let it ring into the answering machine; if she said, "Ben, bring the phone," he grasped the receiver sideways in his mouth and carried it to her. If she was on the phone and needed something on the floor—a dropped sock or a hand towel—she could point to it and Ben would retrieve it. Typical of the breed, he was a one-person dog: Ben bonded with Karen for life.

Karen applied for the job of manager of a day care center for cognitively challenged adults, and she was hired. After a few months of commuting—in her wheelchair and her wheelchair-accessible van—she asked about a small cabin on the grounds that stood vacant and the director said she was free to move into it. She said goodbye to Alice, who had transformed her life by adding a dog to it; they would remain friends for the rest of the older woman's life.

Now Karen and Ben had a home of their own. As Andrew Solomon has written: "The opposite of depression is not happiness, but vitality." Karen was feeling seriously alive.

CHAPTER 8

Logan

s often happens in families whose children wind up on the autism spectrum, the baby started out fine, a "neuro-typical" child. "Logan was so cute!" Donna Erickson told me. "He knew all our names. When Ryder played the guitar, he would dance and say, ' 'Ida, 'Ida!' When Jeff put out birdseed, Logan would run to the window and cry, 'Birdie!' I used to bring him to work with me at the airport and he loved meeting everyone."

"It didn't happen all at once," Jeff said. "Over a few weeks' time—there had been a few medical episodes in a row, including a bad cold, a well-child visit from a circuit-riding nurse, shots, and a scary-high fever—we don't know what triggered it—he stopped talking and making eye contact and he started hand-flapping and spinning in circles."

"Suddenly I couldn't take him to work anymore," Donna said. "He was too weird, spinning and hand-flapping. He stopped sleeping at night. He started running into walls."

Jeff said, "One day a visiting specialist at the village clinic asked me, 'Can I do a couple of things with your son and watch him for a while? Would that be okay?' Afterward she said, 'Have you ever considered having him diagnosed for autism?' "

Jeff phoned Donna at the airport and she rushed to the clinic, asking, "What's autism? What's wrong with Logan?"

"We didn't have a clue," she told me. "We had to look it up. They told us to take him to Nome. The doctor there was very kind—Dr. Kirk Scofield. I was completely exhausted, I hadn't slept for . . . I don't know how long. Dr. Scofield said, 'We'll take care of him, you need to sleep.' When I woke up, he said, 'I need to explain to you what's going on with your son.' "

As the Ericksons drove home, with Logan in the backseat, they felt heavyhearted and lonely. They correctly surmised that they were steering straight into brutally difficult, inhumanly exhausting years. No experts could answer the Ericksons' central question: *What happened to him?* Nor could anyone suggest where they might find help, especially not in their remote and tiny village.

SINCE MEN FIRST DROVE STAKES into the Bering Sea coast two thousand years ago—trying to hammer in a landing point, a whaling camp—the wind has tried to blow them away. The wind had already banished trees and flowers from the water's edge, polished thick sheets of land-, ocean-, and river-ice in every direction, and allowed only a gnarly tundra to creep across the ground in brief summer. For silent eons, ice crystals sparkled up at icy stars. Musk oxen, moose, caribou, reindeer, Arctic foxes, and wolves crunched across the tundra, heads down against the gale. Walrus, bearded seals, ringed seals, gray whales and bowhead whales cruised and plunged through ice-rimmed seas. When the ancestors of the Iñupiat Eskimos roped together antlers, bones, hides, and sealskins for shelter and called the place Unalakleet, "Place Where the East Wind Blows," dogs were with them.

Without dogs, humans could not have sledded to this icy coast on the Norton Sound, nor made contact with other humans once they got here. Elsewhere on the planet, hunter-gatherers and agriculturalists relied on horses, camels, elephants, burros, yaks, and water buffaloes, but those who crossed frozen lands wanted and needed only dogs. Without dogs, the snowy wilderness would have remained untracked, uncharted, and unsettled, until the invention of the bush plane and the snowmobile.

Today the people zip in and out of Unalakleet by air. The one-room airport is as scuffed and homey as the church social hall a few blocks away. There are no security lines or X-ray machines, but there is free coffee and a small lending library. A pilot will linger if a local has forgotten something and asks to run home for it. The coastal villagers still keep dogs. They hunt, trap, and ice-fish with dogs; and they enjoy dogs of all sizes, mostly mutts, some as big and heavily maned as lions, others so small and fluffy they would not look out of place in Manhattan. Dogs trot off-leash along the dozen short roads in the morning and local citizens greet them by name—*Hello there, Bear,* or *There you go, Miki*—as the dogs saunter by. Dogs walk little kids to school; at the schoolhouse door, the children unsentimentally say, "Go home," and the dogs run home.

People who aren't required to be outdoors (average winter temperatures: 11 to –40 degrees Fahrenheit) would just as soon stay inside, but their furry dogs scratch and whine at the doors and shake their jingly collars. Zero degrees is refreshing to a double-coated Alaska dog, and –20 degrees is just about perfect. So villagers "run" their dogs by truck. They drive to the edge of town where the road encircles a barren stretch called the Kouwegok Slough (which, if you say it aloud, gives an inkling of the sound of Iñupiat). They open a car door, let a dog out, and then putt-putt behind as the dog gallops along the iced-over pavement or darts into the snow-thick tundra. When two or more locals are out running their dogs, they laugh at themselves and wave to one another through smudged truck windows.

THE MIDNIGHT SUN HOVERS EYE-TO-EYE with the Ericksons' kitchen window on summer nights, streaking it in radiant orange; in the yard, a rowboat, a snowmobile, aluminum ladders, chains, and empty propane cylinders glow like copper pots. Unalakleet is a working fishing village: massive steel shipping containers sit randomly in lots all over town amid the detritus of engines and boat parts and gasoline pumps, giving the town the utilitarian look of a depot. But there is softness in the hooded, nearly hidden faces of the people and in the parkas, mittens, baby carriers, and boots they wear, stitched of sealskin, moose-hide, rabbit and caribou fur. There is soft light in the Ericksons' small living room lit by wood-burning stoves, where homemade throw rugs and blankets lie like pelts upon the furniture and floors. Some of them *are* pelts.

After Logan's diagnosis, Jeff and Donna—the first parents in town to confront severe autism—studied websites and books. Within the year, they could have *written* the websites and books. Their jolly, round-tummied, happy, chatty Logan was gone. In his place lived an almost-feral boy who growled, flinched, and shrieked, wasn't toilet trained, and did not speak. Handsome like his big brothers, he was also gaunt, pale, and hypervigilant. He had a narrow face, a shock of straight brown hair, and long, elegant hands. His family was athletic, but Logan's own wiry strength and speed were in the service of flight, hiding, rocking, staying awake around the clock, flipping out, and trashing the house in the small hours of the morning. "It's like he has super-human strength," Donna told me.

"Talon was a toddler when Logan was born, and we couldn't be the parents he needed. We had to give Talon to Jeff's sister, who lives up the hill, to raise for us."

"Everything we loved to do—go upriver and camp and fish, go to the boys' basketball tournaments—we had to stop; we couldn't go together anymore," Jeff said. "One of us had to stay home with Logan. Even just hanging out at home, we had to be hawk-eyed. Our neighbors and friends were wonderful—everyone always looked out for Logan, but finally it was on us. Autism took over our life.

"We live two hundred feet from the Bering Sea and Logan's a runner, and a run-to-the-water-er. As vigilant as you think you are—you're watching him, watching him—poof, he's gone."

One Mother's Day, Jeff took Logan outside to jump on the trampoline while he worked on a boat motor nearby and Donna tried to take a nap upstairs. Austen, then a teenager, happened to step out on the second-floor balcony. Gazing across the Norton Sound, he saw Logan's hat floating on the dark water in a break at the edge of the ice. Looking harder, he saw Logan's face, cupped by the surface of the water. Logan wasn't struggling or yelling for help—this fact struck them all later—he was just moaning very faintly while sinking. Austen screamed for his dad in the yard below him. A visiting friend of Austen's was the first to reach Logan. By then Logan had slipped between two layers of ice and was submerged in 30-degree water. Only his nose, lips, and forehead remained visible. In another minute, he'd have disappeared. They were doing their best, the parents knew; they trudged through life exhausted and despairing, trying their hardest, but it wasn't good enough.

EVERY MARCH, DOG-LOVING UNALAKLEET IS a checkpoint along the Iditarod, the 1,100-mile wilderness endurance sled-dog race between Anchorage and Nome. Beside the frozen Unalakleet River, a two-story aluminum-sided building is turned over to the race. Volunteers and sports fans, including veterinarians, come from all over the world to help out. As the dog teams glide into the checkpoint (beckoned, at night, by fires and village lights visible two miles out), the drivers and the dogs are welcomed like old friends. Many of the mushers *are* old friends of the locals, and some of the dogs are old friends, too, recognized, stroked, and addressed by name. The dogs are attended to first, closely inspected limb by limb, paw-pad by paw-pad, toe by toe, toenail by toenail, for any sore or weakness at this stage of the ten- or eleven-day race, then fed hot meat stews, bedded down on fresh straw, and covered with blankets. They curl up and deeply sleep, basking in sunlight by day or under wild starlight at night. After bedding down the dog team, the dazed, frosted, filthy, snow-burnt mushers—young and middle-aged men and women—limp up a steep ramp of snow and into the muddy-floored race headquarters. Along a side wall, curtained-off cells offer narrow cots and blankets for mushers who need to collapse for a while. Folding tables and chairs are set up for those who want to eat, while apron-wearing volunteers cook over double burners in the back of the room. A GPS-equipped computer screen broadcasts the location of every team in the race. The racers sign in, use the facilities, and accept Styrofoam cups of hot coffee and paper plates of bacon and eggs, or of barbecued meat. Some don't talk or look around much; others briefly relax and regale the volunteers with tales of the road before hobbling back down to their dog teams and shoving off for the next checkpoint.

DEEDEE JONROWE, IN HER EARLY SIXTIES, looks as glamorous as it is humanly possible to look when living around-the-clock with dogs in the subarctic wilderness. In a neon-pink down coat, leaning into a turn with wind-tossed blond-and-silver hair under a crystal-blue sky, she offers a scene of startling beauty. The most decorated woman still racing in the Iditarod, DeeDee has the tawny tan and laugh lines of a veteran winter outdoorswoman. She and her husband live in Willow, Alaska, with sixty-five dogs, of whom sixty-two are sled dogs, two are yellow Labs, and one, Mr. Myagi, is a Pekingese. A three-time runner-up, sixteen-time top-ten finisher, and 2013 inductee into the Mushers Hall of Fame, in 2003 she ran the Iditarod three weeks after completing chemotherapy for breast cancer. A favorite of fans, fellow mushers, and veterinarians, she's a multiple winner of the Sportsmanship Award, the Humanitarian Award, the Most Inspirational Musher Award, and the Best Cared-For Team Award. In the off-season, DeeDee stays in shape by running half-marathons.

She raises Alaskan huskies, working dogs of mixed backgrounds bred for speed and endurance. "I have to be careful not to give my dogs too many clues that it's almost Iditarod time," she told me. "Out of dozens of dogs, I can only take fourteen every year. When they smell the harnesses or see the sled come out, they go nuts! They ramp up. They run around like, *Pick me! Pick me!* If they didn't love to run, they could easily go into their doghouses and poke out their faces and just look at me; that would tell me they'd rather not go. I would never force a dog to go! But that's not what they do. Ever.

"I have a degree in genetics," she told me. "Breeding toward a particular type of success is second nature to me. I work with a blend of breeds. My dogs are born to be sled dogs. They're not all necessarily gifted enough to be top race dogs, they may not be built for speed, but they're born to run.

"I keep entire litters. I just retired one litter at eleven years old. They spent their entire racing careers with me and are now mentoring the young dogs. I have a senior center for my dogs ten years of age and older. My dogs usually live to fifteen or sixteen. I play with them, I sit with them. I think about their heydays. I remember our finishes together, when *they* were the stars, the macho-men or the girls of the team. I think about the storylines we have together. For my old dogs, it's all about making them feel their lives were successful and that they have earned this quality of life.

"We do vet-check for all the dogs lined up to race each year. The other day, I took along Shakespeare. He's thirteen now, a really fine leader in his day. I think he thought it was pretty cool to ride in the truck with the young racers, but he's all about retirement. He came along for the ride, but he was very chill. He wasn't planning to race. Sometimes I put a harness on him so he'll show the young ones how it's done, but there are days he looks at me like, *You know, I'm not totally into this today*. So I take it off. I tell him, 'That's fine, you don't have to mentor the young ones. You're my pet. You have nothing to prove, Old Man, you are just my love.'

"In the early days, people couldn't have traveled ten miles in Alaska without dogs. They were a powerful influence in Native culture. The old people remember when there were no snowmobiles or cars out here; it was all about the dogs. Then people discovered you could buy a snowmobile and leave it lying outside and never have to feed it or pay attention to it and the culture began to shift. The Iditarod revived the dog culture and gave economic value to the dogs again. My dogs carry the bloodlines of the original village dogs, with added blends of the best of the best over the years. My dogs are born to pull. They may not all be great racers, but they love to pull."

THE IDITAROD DOGS ARE WIDELY REGARDED as the most athletic dogs in the world; Alaskans are proud of them, remembering the names and achievements of the great ones. "These are the best-cared-for dogs I've ever seen," Kathy Fauth, DVM, a Chicago-area veterinarian, told me during a lull at the checkpoint. "There are animal lovers who criticize the Iditarod and I appreciate their concerns, but most of the household pets in my practice back home don't get this level of care, attention, and knowledge, or anything like this constant companionship and activity and opportunities to socialize. These are working dogs. They're treated like elite athletes and they're loved. The mushers cry—male mushers cry—when they have to 'drop' a dog—leave it behind with us to be flown home. The dogs who are left in our care absolutely know when their teams, in the distance, are taking off from Unalakleet and they howl and cry."

Vern Otte, DVM, of Leawood, Kansas, told me, "We vets will refer to a dog by his number—'Bring me 35C'—and the musher will say, 'You mean Frosty?' They're so careful with the dogs. When 'dropping' a dog with us, they'll say, 'She's tired' or 'He doesn't want to run' or 'He's not having fun.'" (At this, the doctor's eyes filled with tears.) "If the dogs don't want to run or to work, they don't have to. Dallas Seavey, a young man, late twenties, pulled in with one of his dogs wrapped in a blanket, all tucked into a basket on the sled. 'Is he hurt?' I asked him, and Dallas said, 'No, not at all, I just thought he might enjoy a break.'" (Seavey would win the 2012, 2014, 2015, and 2016 Iditarods.)

"The dogs are magnificent," said Dr. Otte. "They trot in, looking at all the people, enjoying the attention. They sniff the straw to see which dogs have come through already. They greet the other dogs. If they're staying, they seem to know it. They eat, curl up, and sleep. If they're not staying long, they know that, too: their tails are wagging, they're barking and barking, like *Not tired! Feeling good! Let's go, let's go!*"

The winningest dogs these days are the offspring of Arctic breeds like the Eskimo dog and the Siberian husky, *crossed* with greyhound, German shorthaired pointer, Irish setter, or other hounds or gundogs. They're leaner than malamutes and Siberian huskies, and many look as sweet, skinny, and hopeful as shelter mutts. One ear up, one ear down, brown and white patches, a white stripe down the nose. Many have blue eyes—or one blue eye—an inheritance from a Siberian ancestor.

A few Iditarod mushers still race with all-Siberian husky teams—for their history and their beauty—and the arrival of an all-Siberian team generates excitement in the villages and the rapid whirr of digital cameras, but the all-Siberian teams don't win the Iditarod. The champion dogs of the modern Iditarod—well, you'd miss them if they ran past you on the sidewalk. You can't tell their greatness by looking at them.

WHILE FIELD BIOLOGISTS AND SCIENTISTS in the rest of the world declined to study dogs, dismissing them as invented commodities, Alaskans didn't make that mistake. Alaskans, who knew their state would never have been discovered or settled without the sled dogs, never asked themselves whether their canine partners were real animals or not.

But it *is* true that between the Wolfdog of the Chauvet Cave in southern France and the chubby, hopeful rat terrier staring with unblinking intensity at your peanut butter sandwich, a genetic bottleneck occurred, a tremendous loss of genetic diversity. It happened in the time of Queen Victoria and it would cost dogs the respect of many decent animal-loving people for a very long time.

EIGHTEEN-YEAR-OLD VICTORIA, who ascended to the throne in 1837, enjoyed keeping a lively pack of dogs about her person at all times. Soon simply *everyone* had to have a pet dog. "For a fashionable woman in Victorian England a pet miniature dog was as indispensable as an opera box or presentation at court," wrote a *Harper's Bazaar* columnist in 1893. "She was nobody without her pet who accompanied her wherever she went and was fed and housed . . . as daintily as the heir to the title and estates." "Nobody who is anybody can afford to be followed about by a mongrel dog," warned a dog breeder in 1896, part of a new hierarchy of specialist breeders governed by the newly founded Kennel Club.

The wealth of empire and industry gave British aristocrats the time and money to dabble in the new hobby of "dog fancy," and an entertaining race was on to create and perfect coveted breeds. "Somebody would show up on the promenade with an exotic-looking dog, and then label it with a breed name: 'Check out my new Dalmatian!' " writes British dog-fancier and historian David Hancock. "Other owners of these breeds would say: 'Wait a minute, there's no way you can call that thing a Dalmatian, this is a Dalmatian.' "

Must-have dogs of the day included "purebred" Cavalier King Charles spaniels (the queen's favorites), Skye terriers, Yorkshire terriers, Sealyham terriers, wire and smooth fox terriers, bulldogs, schipperkes, Pekingese, Airedale terriers, English fox hounds, Dalmatians, Gordon setters, Italian greyhounds, Shetland sheepdogs, Scottish terriers, West Highland white terriers, cairn terriers, and Pomeranians.

How did Victorian dog breeders come up with all these types so quickly?

Despite later confusion, they did not invent them ex nihilo.

"BY TEN THOUSAND YEARS AGO, dog-keeping . . . had spread throughout much of Europe, Asia, Africa, and the Americas," writes Dr. John Bradshaw. "Soon after this, and in many parts of the world, recognizably distinct types of dog appear." There was a deliberate element in the breeding, he says, through "the simple expedient of allowing bitches to mate only with chosen males of similar type." But dogs continued to choose their own mates, too, leading to healthy diversity.

Mastiff-types evolved to guard villages and houses, sight hounds and scent hounds to track and hunt, terriers to eradicate vermin and run small game to ground, sheep-herding dogs to protect or to herd livestock, sled dogs for transport in snowy lands, and even (in ancient Rome and China) lapdogs for the amusement of the elites.

The breeding programs of the nineteenth-century British leisure classes took these local varieties, or "landraces," as their raw material. They weren't selecting for intelligence, speed, or skill, but for increasingly arbitrary appearances. "They were selecting dogs that would turn heads, inspire envy, and look best as familiars in high-class portraiture," writes economist Rosie Cima. "All of a sudden, thousands of people were breeding dogs as ornamental luxury items."

"The Victorians," writes University of Minnesota professor Andrzej Piotrowski, "made a dog show into a celebration of their own ability to alter species." It's no less true today, with disastrous consequences for many cherished types of dogs who have been inbred to the point of collapse.

BECAUSE MOST MODERN DOG BREEDS arose in the past 150 years, and because their origin stories are entwined with gentlemen in tweed and top hats, and ladies in silk, ribbons, and lace, it began to seem as if dogs were just likable contraptions invented or upgraded in the years prior to the Great War. If a dog today isn't recognizable as a purebred, then we assume he's a Victorian hodgepodge, as if someone took the handset from the telephone, the crank-handle off the motorcar, and the lightbulb from the electric lamp and made a funny new thing.

Dr. Ray Coppinger recalls searching, as a child, through picture books of dogs to determine the provenance of his sidekick, "Smoky," a little short-haired black mutt, the gift of his uncle. "Uncle Joe and I finally decided Smoky was a crossbred Italian greyhound," he writes. "That didn't seem improbable at the time, and he had to be some mixture, didn't he? Plus, it gave me an impressive answer to the frequent question, 'What kind of dog is he?'"

The novelist Ann Patchett describes a similar wonderment regarding the ancestry of her cherished "Rose," a happy little white dog with moist black eyes and upright ears, rescued from a parking lot. "She has been, depending on how one holds her in the light, a small Jack Russell, a large Chihuahua, a rat terrier, a fox terrier, and a Corgi with legs," Patchett writes. "Currently she is a Portuguese Podengo, a dog that to the best of my knowledge was previously unknown in Tennessee. It is the picture she most closely resembles in our *Encyclopedia of Dogs*. We now say things like, 'Where is the Podengo?' and 'Has the Podengo been outside yet?' to give her a sense of heritage. But really, she is a parking lot dog, dropped off in a snowstorm to meet her fate."

I won't pretend to have resisted the notion that every mutt descends in some part from British nobility, not unlike an amateur genealogist discovering a family tree rife with dukes and duchesses rather than with peasants and serfs. We sent in a DNA swab to WisdomPanel.com from Henry, our small tan wirehaired mix who best meets the description "a funny-looking little dog." Rescued in puppyhood from a shelter, his approach to new people is at once super-hopeful and abashed. He rushes in to deliver some kisses to the new person, then curls in on himself and sidles away, then charges in to give more licks, then sidles away again.

"CONGRATULATIONS!" replied Wisdom Panel in all caps several weeks later, when delivering Henry's genetic results, or "Insights." They did not disappoint. The scruffy twelve-pounder, they said, descends from the Cavalier King Charles spaniel (Queen Victoria's favorite!), the Chihuahua, the Irish setter, the shih tzu, and—to lesser degrees—the Saint Bernard, the Norfolk terrier, the German spitz, the Yorkshire terrier, and the Small Münsterländer.

Naturally we refer to him as the Small Münsterländer. As in, "Can someone go open the door for the Small Münsterländer?"

BUT TO DESCRIBE what an unidentifiable mutt might be—if *not* a mélange of purebreds listed in Kennel Club studbooks—would require a knowledge of deep time. To guess at the antecedents of happy little mutts like Smoky, Rose, and Henry is to tumble backward in time before the first greyhounds, Podengos, and Münsterländers walked the earth. It is to catch a glimpse of what twentieth-century scientists said did not exist: a biological niche for the domestic dog.

To gain respect for *Canis lupus familiaris* as an authentic animal, one could start 60 million years ago with the world's first carnivorous mammals; or 50 million years ago when the order Carnivora split into (dog-like) caniforms and (cat-like) feliforms; or 40 million years ago with the birth of the *Canidae* family in North America, the forerunners of wolves, jackals, foxes, dingoes, and dogs.

You could start 10 to 8 million years ago, when a variety of canid hunters wandered from North America across the Bering Strait land bridge onto the Mammoth Steppe of Eurasia and joined "the rich palette of predators and scavengers that co-evolved with herding ungulates." You could zoom in on Eurasia 4 million years ago: "a vast playground for *Canis* evolution," where the canids preyed upon "an array of creatures more bizarre than the imagination could invent—giant woolly forms with grotesque antlers, horns or tusks, long manes, humps, and trailing tresses . . . the mysterious fauna on the cave walls." One biologist describes that world as "Where the Wild Things Were." Two million years ago, canid dominance in Ice Age Eurasia was so commanding that paleontologists refer to the period as "the Wolf Event."

One could start the story 800,000 years ago with the appearance of the greatest of mammalian predators: *Canis lupus,* the gray wolf, the last common ancestor of the modern gray wolf and the dog. For hundreds of thousands of years, gray wolves were the apex predators in the Northern Hemisphere: they hunted as a pack and shared the kill, mated monogamously, lived in family groups, and offered long childhoods to their pups while expecting their adolescent offspring to stay with the pack and help babysit. "All members share food and parental care generously," write paleontologists Wolfgang M. Schleidt and Michael D. Shalter. "Even siblings and friends share food and affection . . . This cooperation and risk-sharing not only among close relatives, but among individuals bonded as mated pairs or by lasting friendships among individuals of the same gender, is the central feature of canid pack living."

Then ten, twenty, or thirty thousand years ago, the gray wolves encountered anatomically modern human hunter-gatherers emerging from Africa on the trail of the same megafauna. They, too, were becoming pack hunters. The two species became fellow travelers and the domestication event or events—or the start of a mutual coevolution—began.

The magnificent gray wolves of prehistory—famous for their night songs, their group hunting strategies, their daring assaults, their close-knit families—*these* animals, and not only the adorable lapdogs and sleek gundogs of Victorian England, are the ancestors of all modern dogs. The genetic bottleneck created by Victorian dog-fanciers is a recent and minor footnote in dog history; the vast majority of the *billion* dogs alive today—the village dogs, pariah dogs, feral dogs, free-ranging dogs—do *not* carry that asterisk. The true origin story of the dog is still being written, but in *that* story, dogs are clearly something other than market-driven merchandise, "intentional creations of human ingenuity."

While that deep story was inaccessible, dog studies were considered absurd, as if a scientist were too lazy to strap on a knapsack, take off in a bush plane, and go find some *real* animals.

TODAY THE IDITAROD PAYS TRIBUTE to the great, pioneering Arctic sled dogs: thick of fur, curly of tail, regal of bearing, often blue-eyed. Wolfish-looking, powerful, they *appear* to have missed the Victorian creation-of-purebreds event. And so they did.

The first sled dogs likely evolved in Mongolia thirty thousand to fifteen thousand years ago. (Actually, the latest genetic analysis of modern dogs zeroes in on central Asia, including Mongolia and Nepal, as the possible place of origin of *all* dogs, adding central Asia to the atlas of dog evolution that already pinpoints Europe, the Near East, Siberia, and China as possible sites of dog domestication.)

Asiatic people wandered with their dogs across the Beringia land bridge from Siberia into Alaska twenty thousand to fourteen thousand years ago. Their descendants—both humans and dogs—radiated southward and populated the Americas. Those dogs evolved into a variety of landraces, including the "Indian dogs" of the lower forty-eight states, the ginger-colored "yaller dogs" or "Carolina dogs" of the southeastern United States, the Chihuahuas and Xoloitzcuintlis (hairless dogs of Mexico), and the *perro sín pelo del Peru* (hairless dogs of Peru).

Thousands of years later, Columbus landed and the European Conquest began. The Europeans brought dogs with them, including the Spanish Conquistadors' "dogs of war," deerhound/mastiff mixes "festooned ... with padded armour and spiked collars" who terrified and brutalized the indigenous populations.

American dogs perished in the Conquest. They died in battle and they were wiped out by infection from European pathogens (which decimated the human populations, too). The surviving indigenous dogs mated with the European dogs until they were swallowed up in the new American melting pot.

It has long been accepted that the indigenous landraces of

American dogs—from the Arctic dogs to the Carolina dogs to the Indian dogs to the Chihuahuas—went extinct. It has been assumed that modern North and South American dogs trace their family trees back to Europe and the Western monarchies, that they descend from the dogs of the European conquerors and colonists (who came by boat across the Atlantic Ocean hundreds of years ago) rather than from the dogs of the original Asiatic settlers of the New World (who came by foot and dog-sled across the Bering Strait land-bridge tens of thousands of years earlier).

But some of these pre-Columbian landraces have passionate admirers who claim that their favorites are alive and well. On the southern coast, fans of the Carolina dog call them "old yaller dogs," "Dixie dingoes," and "America's natural dogs." The mushers of the Iditarod race with dogs who descend, they claim, from the original Mongolian sled dogs. DeeDee Jonrowe says: "My dogs carry the bloodlines of our original village dogs, the dogs who carried the mail and hauled the supplies." Alaskans accept her declarations matter-of-factly. Lots of mushers claim ancient origins for their dogs, if not descent from *wolves*. Only outsiders see this as romanticizing.

Wondering whether any indigenous American breeds *had* survived the European Conquest, a research team with the KTH Royal Institute of Technology in Stockholm collected mitochondrial DNA from a wide range of American breeds. The results astonished them: "Many modern North American dogs," reported lead author Dr. Peter Savolainen, "continue to carry a significant genetic signature from their distant past." The Carolina dogs were among those found to have significantly high indigenous ancestry. The Chihuahua has a stretch of DNA identical to that of a dog living a thousand years ago. And the investigators found *no European influence at all* for the Arctic, Inuit, Eskimo, and Greenland dogs.

In a more recent study, researchers with the Swedish Museum of Natural History found that Siberian huskies and Greenland sled dogs "inherited a portion of their genes from the Taimyr wolf," which lived in Siberia 35,000 years ago, giving support to the theory that dogs may have been domesticated as early as 30,000 years ago.

Thus, from Alaska to the Carolinas to South America, there are dogs among us whose ancestors skipped the Victorian breed-creation event and came to the New World on foot, via Asia, tens of thousands of years ago.

As modern Alaskans pay tribute to ancient sled dogs, their wilderness race likely includes direct descendants of the world's earliest sled dogs. When a modern plague seeped into this Eskimo village on the Norton Sound, it was the mushers of the Iditarod who thought about turning again—as in the Old Way—to dogs for help.

CHAPTER 9

Hero Dogs

L ike many dogs, Karen Shirk's Ben seemed to sense
something different about children; he was gentle and
forbearing with them. But one day at a park, tearing
around off-leash with other dogs in a field, he suddenly broke

away from the pack and dashed to a shaded, wood-chip-covered playground area, in the center of which stood a tall metal swing-set where a little girl dreamily glided forward and back. Ben stood directly in the swing's path, facing the child, and barked at her. He dodged at the last minute as she pumped forward and continued barking as she rode backward on her arc.

"Ben! No!" cried Karen from the wheelchair. She'd never had to scold him like this before! "Ben! Come!"

As the child swooshed within a few inches of Ben's face, he turned his head sideways and bared his teeth, as if to grab a bite. Karen came wheeling in from the left as best as she could maneuver over wood chips and the child's mother ran over from the right, but neither was fast enough. Ben snagged the child's pants leg and yanked her off the swing at its lowest point. She landed her bottom on the wood chips, sat stunned for a second, and then began to wail.

Oh my God, he's lost his mind! Karen was thinking. *He's sick. He must be sick.* She pulled up in the wheelchair, grabbed his collar, and attached the leash. "I'm so sorry! I don't understand it. He's never done anything like this before. Ben, oh my God." The mother picked up her crying daughter and ran in the other direction, then laid her out on a park bench to examine her for bite marks.

A blasé little boy entered the scene, took a seat on the swing, and gave himself a push-off. Ben tried to wrestle away from Karen's grasp, but she held on tightly. Ben barked at the boy from afar. "What on earth has gotten into you?!" Karen cried. From the far side of the playground, the little girl's mother was looking over reproachfully. Between them on the swing-set, the little boy reached the swing's crescendo when one chain came loose and dropped the child from midair. He hit the wood chips with a grunt and began screaming, and *his* mother ran over.

Karen and the little girl's mother looked at the boy, then at each other, then at Ben.

"Oh my God. How did he know?" cried the girl's mother.

"I have no idea," Karen yelled back.

"How did Ben know that swing was about to break?" she asks now. "Who knows? In the end, some things about dogs are just mysteries; it's like they can do magic."

THEN HE PERFORMED A BIT of magic for Karen.

She underwent heart surgery in 1998 and returned home to her cabin on the grounds of the adult day care center to recuperate. Her insurance covered nursing care forty hours a week, but left her alone at night. One night, the morphine pump malfunctioned and knocked Karen instantly into a deep sleep. The IV drugs ticked into her system too quickly, in a deadly combination of pain-killers. Her heart rate slowed, her breathing slowed and became labored, her blood pressure fell . . . she dropped from sleep into unconsciousness . . . her body temperature fell . . . she plummeted toward shock and beyond, into a zone of grave danger.

Around 11:00 p.m. that night, the phone rang into the silent cabin. Presumably Ben waited—as he'd been trained—for Karen's command, "Bring the phone." No command came, so his job was to let the phone ring into the answering machine, but that night—these facts are undisputed—Ben picked up the receiver, dropped it on the bed, and barked and barked and barked. It was Karen's father calling to check on her.

"The phone picked up," says Captain David Shirk, "and Ben started to bark. He barked incessantly. Now that's not like Ben, he's not a barker. I yelled, 'Karen! Karen, can you hear me? Karen, what's wrong?' When he would stop barking, I could hear very heavy, labored breathing. It sounded like there was something seriously wrong. I knew Ben was upset. I had to trust his judgment that there was a problem." Captain Shirk hung up and called 911. The rescue team, who found the phone off the hook, told Karen she wouldn't have lived through the night.

TALK ABOUT LASSIE! Did this really happen? *Can* these kinds of things really happen?

They happen all the time—in popular culture. Heartwarming neighborhood, city, national, and international news reports pay tribute to dogs saving people and other animals in peril. Rescued dogs smell smoke and evacuate their families; on a walk, they drag their owners down the block into a stranger's backyard where an elderly man has been pinned under a fallen branch for ten hours. Headlines trumpet their achievements: "Tiny Dog Rescues Girl from Attempted Abduction" (Yahoo News). "Dog Finds a Tiny Kitten, Risks Everything to Save Her" (Fox News). "Dog Saves Baby from Abusive Babysitter" (ABC News). "Dog Tries to Save Dying Fish" (YouTube, Phetchaburi, Thailand). "3-Year-Old Siberian Girl Discovered after 11 Days Lost in Wilderness When Her Trusty Dog Summons Rescuers" (*Siberian Times*). "Stray Dog Brings Home Newborn Baby Wrapped in Plastic and Thrown in Trash Dump" (Khaosod Online, Thailand). "Heroic Senior Dog Saves Her Family from Charging Moose" (*Boulder County News*). (I am not making these up.) "Hero Pit Bull Shows Up Lassie, Uses iPhone to Call 911 and Save Owner's Life" (HuffingtonPost). (I couldn't even make these up.)

Such stories surround us, breathlessly shared by one and all. More *common* occurrences—Owner Falls Down Stairs, Dog Keeps Napping—or Family Wakes Up Their Sleeping Rescue Dog and Carries Him Out of House to Escape Fire—go unreported.

Are the hero tales exaggerated? It would be impossible to separate fact from fiction until someone finally paid serious scientific attention to dogs.

IN 1973, ALAN BECK, a curly-haired, bearded young man from Brooklyn, was pursuing a doctorate in animal ecology at Johns Hopkins University when Professor Edwin Gould of the School of Public Health suggested he consider urban dogs as his subject (rather than prairie plants).

Oddly enough, despite the fact that feral, stray, and free-ranging dogs roamed most of the villages and cities of the world, the few surviving *wolves* on earth had been studied far more extensively. In fact, there was no record of *anyone* having studied urban dogs. Alan was game. "So I studied the dogs of Baltimore and pretended they were wolves," he says. He tracked free-ranging dogs around the city, "driving not a Land Rover but a used sedan." By car or on foot, Beck hurried after lone dogs and pairs of dogs and packs of dogs at all hours of the day and night, carrying a camera and cassette tape-recorder; sometimes he rode along with city rat control crews. "On different occasions I was asked if I was a sanitation inspector, dog catcher, newspaper photographer, or narcotics police officer," he writes.

"In the early morning hours before and after sunrise they are most easily seen," he reported for *The New York Times* about that city's packs. "They move along in packs of six to eight, cadging scraps of food from garbage cans, dodging cars as traffic begins to build up in the wakening streets, then crawling into abandoned factories or garages for shelter."

He learned many things about city street dogs: their tendency to roam in packs of at least two; their preference, during the summer, to forage at dawn and sunset to avoid midday heat; and their frequent interactions with the edges of civilization. Even feral dogs living in pockets of forest around Baltimore emerged to visit alleys, knock over garbage cans, and rummage through the remains, and some accepted regular handouts from human acquaintances.

The Ecology of Stray Dogs: A Study of Free-Ranging Urban Animals was published in 1973 and would become a classic in urban studies. Overnight, Alan Beck became the world authority on urban dogs, simply, he says, by being the only one to study them. In one keen observation, he wrote: "It is obvious that man is very much part of the ecology of the dog."

Decades would pass before the significance of Beck's observation—that dogs have an "ecology," a habitat—began to be unpacked by biologists. Coincidentally, when the breakthrough occurred—what Dr. Ádám Miklósi would call "the rediscovery of the dog"—it would happen almost simultaneously on two continents, starting with almost-identical discoveries.

AROUND THE TIME THAT BEN was trying to drag a little girl off a playground swing in Ohio, Professor Vilmos Csányi, chair of the Department of Ethology at Eötvös Loránd University in Budapest, was speculating about whether "hero dog" anecdotes and other dog lore might serve as a starting-point for scientific inquiry. Scientists still snubbed dogs as research subjects, while strenuously avoiding the epic literature about amazing canine acts of intelligence and loyalty, about dogs waiting for years for their masters to return, standing guard over their masters' graves, and rescuing imperiled children because storytellers often ramp up the facts of a case.

Scientists naturally discounted such wonder tales—fictional

ones and those posing as nonfiction—as if the very existence of Lassie Myth-type stories ruled out the possibility of their containing shards of truth.

But Dr. Csányi—whose lab had focused for several years on the learning processes of the East Asian labyrinth fish—loved dogs, and wondered if there were any tidbits within the vast "oral history of dogs, of anecdotes and beliefs," that scientists might confirm or disprove.

Everyone had heard of Balto, the Siberian husky who led his sled-dog team through a blizzard to the remote outpost of Nome, Alaska, in the winter of 1925, delivering diphtheria antitoxin to stem a deadly outbreak of disease. Balto led the last two legs of the 650-mile relay and was the first to enter Nome at 5:30 a.m. on February 2. A statue of Balto was erected in Central Park (he was present at its installation) and the modern Iditarod race commemorates the serum-run, one of the last great achievements of the sled dogs and their mushers, before they were rendered obsolete by bush planes and pilots.

People heard of an Akita Inu named Hachiko, who joyfully accompanied his master, Hidesaburo Ueno, a widely esteemed professor of agricultural engineering at Tokyo University, to and from Shibuya Station every day until an afternoon in May 1925 when Professor Ueno failed to appear. He had suffered a fatal hemorrhage while giving a lecture earlier that day. Hachiko returned to the station the next afternoon to meet his missing master, and the afternoon after that, and the one after that every day for nine years until his own death. In 1935, the community erected a monument to his memory, in which he sits attentively outside Shibuya Station, eternally waiting for the afternoon train. That beloved but bittersweet statue, in time, seemed to prolong the dog's yearning rather than simply honor it. On the

80th anniversary of Hachiko's death, in March 2015, the University of Tokyo unveiled a *new* statue in which, ninety years after Hachiko began waiting, Professor Ueno returns. The dog rears up in greeting, placing his front paws against his master's suit jacket, while Ueno reaches out his hands to ruffle Hachiko's fur, and the pair joyfully scan each other's faces.

Most people knew of Argos, the hero of one of the most poignant dog stories of all time, as told by Homer in the *Odyssey* in the eighth century BCE. After a twenty-year absence, Odysseus has just returned to his home in Ithaca disguised as a beggar. No one recognizes him . . . except his dog. Once the hero's treasured and fearless companion, Argos is now an elderly bag-of-bones whom someone has thrown out with the garbage. Newly arrived, the disguised Odysseus stands talking to an old friend who does not recognize him.

> While he spoke
> an old hound, lying near, pricked up his ears
> and lifted up his muzzle. This was Argos,
> trained as a puppy by Odysseus
> but never taken on a hunt
> before his master sailed for Troy.
> He had grown old in his master's absence . . .

In his prime, Argos had been "swift and strong / he never shrank from any savage thing he'd brought to bay in the deep woods; on the scent, no other dog kept up with him," explains the friend. "Now misery has him in leash. His owner died abroad."

> Treated as rubbish now, he lay at last
> upon a mass of dung before the gates—

> . . . Abandoned there, and half destroyed with flies,
> old Argos lay.
> But when he knew he heard
> Odysseus' voice nearby, he did his best
> to wag his tail, nose down, with flattened ears
> having no strength to move nearer his master.
> And the man looked away, wiping a salt tear from his
> cheek . . .

The supreme effort to wag his tail one last time for his master is the end of the great dog:

> Death and darkness in that instant closed
> the eyes of Argos, who had seen his master,
> Odysseus, after twenty years.

Argos was a fictional dog; still, his loving fidelity has resonated with dog lovers across all recorded time.

In the final years of the twentieth century, a reconsideration of the capacities of dogs, and of the bona fides of dogs as real animals, seemed, to Dr. Csányi, to be long past due. He announced that his lab would be turning away from the study of East Asian labyrinth fish to the study of dogs. As the aquariums were carted out, Ádám Miklósi, his younger colleague, recalls a sinking feeling: turning from fish to dogs "did not seem to be much of an improvement."

THE BUDAPEST GROUP BEGAN BY looking at the apparent ability of dogs to understand finger-pointing and other "human communicative signs." "Pointing is one of the most widely used human nonverbal gestures for indicating objects," writes Miklósi, now director of the Family Dog Project and head of ethology at Eöt-vös Loránd University. "Even superficial observation reveals that humans also use this form of gesturing when interacting with dogs." The ethologists wondered how well dogs really understood pointing (sheepdogs in the field certainly seemed to respond accurately); and, *if* this was a special canine skill, was "the ability to work with humans . . . an important factor at some point in dog evolution?"

MEANWHILE, IN ATLANTA, GEORGIA, PROFESSOR Michael Tomasello of Emory University, a comparative and developmental psychologist and an affiliate at the Yerkes Primate Center, found himself getting tangled up with dogs instead of with his usual subjects, chimpanzees. Also intrigued by pointing, he was trying to find a *chimp* who understood the gesture.

After about nine months of age, a *baby* understands that she can draw someone's attention to an object beyond her reach by gazing or pointing at it; and that when an adult or older child points or gazes at something, it's meant to attract *her* attention to it. The scientific literature calls it "intention-reading" or "communicative intention." "Look at the bluebird! See the bluebird in that tree? Yes, there it is!" we burble at babies; or (when the baby is doing the pointing) "You like that balloon? Yes, isn't it bright and green? No, you can't hold it because you'll bite it and it'll pop and you'll start crying. Yes, I saw it already. I know you want it, you can't have it, okay?" The baby doesn't yet possess speech, but you and the baby are clearly communicating. It turns out that intention-reading is a cornerstone of human communication.

Did our closest primate relatives, the chimpanzees and bonobos, share communicative intent? If *not*, did intention-reading arise *after* our lineage diverged from the last common primate ancestor five to seven million years ago? Was intention-reading *uniquely human*? The Emory researchers, including an undergraduate named Brian Hare, modified an experiment in which a researcher secretly hid a treat under one of two containers, faced a chimpanzee, and pointed to the container concealing the food. The chimpanzee then chose a container and scored a treat, or not.

Almost invariably, the chimpanzees paid *no attention* to the researcher's signaling. Their treat-finding success rate was no better than chance, even if the researcher also *gazed* at the correct container while pointing, or *touched* it. The chimps were clueless that the researchers were trying to relay an important message: *the treat is here!* Dr. Tomasello was moving toward the conclusion that "spontaneously understanding another's intended communication was a kind of genius unique to humans."

Brian Hare, the undergraduate, whose family lived near campus, had spent many happy afternoons at batting practice in his backyard, relying on his dog, Oreo, to retrieve baseballs from behind any bush or tree at which the young man pointed. Now he said to the professor: "I think my dog can do it."

"Sure," the older man replied. "Everybody's dog can do calculus."

But Dr. Tomasello was prevailed upon to watch a video of Brian's dog. When the treat-hidden-under-a-container test was offered to Oreo, he aced it. Then the dogs of local volunteers passed the test with flying colors, and even puppies could follow a pointing finger to the correct container, and it upset everything scientists thought they knew about communicative intention and primates and evolution and dogs.

(Let me go on record now and state that *my* dogs, despite living three blocks from Emory University, do *not* engage in intention-reading with regard to a pointing finger. I can helpfully point at a neon-green tennis ball under a bush, approaching it and gazing at it while pointing, until I am squatting directly above the tennis ball, which the dogs frantically wanted ten seconds earlier, pointing at it from one and one half inches away, and little Henry will bound over and use the opportunity to lick my pointer finger.)

The two university research groups, in Budapest and in Atlanta (apparently working with dogs with higher SAT scores than my dogs), shared in the discovery that most dogs could interpret human gestures like pointing. The discovery that *dogs* understand our communicative intentions, when even *primates* do not, carried with it a deep revelation.

"We have met the enemy, and they are ours," wrote Commodore Perry in 1813, announcing a naval victory over the French on Lake Erie.

"We have met the enemy, and he is us," said Pogo the Possum of the comic strip *Pogo,* sadly surveying a patch of Okefenokee Swamp ruined by garbage and junk.

At the tail end of the twentieth century, two research groups effectively realized: "We have found the dog's natural habitat, and it is us."

TODAY, RESEARCH INTO THE MYRIAD WAYS dogs have exploited their niche beside humanity—and about how humanity has been shaped by their presence in our lives—is booming, and at prestigious universities. The Duke Canine Cognition Center (founded by Associate Professor of Evolutionary Anthropology Brian Hare); the Canine Cognition Center at Yale; Barnard College (where bestselling author of *Inside of a Dog* Alexandra Horowitz is a psychology professor); Emory (where Dr. Gregory Berns, a professor of neuroeconomics, is scanning the brains of willing dogs in MRI machines); Purdue University (where Dr. Alan Beck directs the Center for the Human-Animal Bond); Hampshire College (where Dr. Ray Coppinger is professor emeritus of biology); University of Colorado, Boulder (where Dr. Mark Beckoff teaches ecology and evolutionary biology); and the Canine Science Collaboratory at Arizona State are making headline-worthy discoveries on a monthly basis. Abroad, centers of study include the Family Dog Project at Eötvös Loránd University (founded in 1994, the first research group in the world to focus on the dog-human relationship); the Clever Dog Lab in Vienna; the Anthrozoology

Research Group in Victoria, Australia; the Max Planck Institute for Evolutionary Anthropology in Leipzig, Germany (where Dr. Tomasello is now codirector); the Animal Welfare and Behaviour research group at the University of Bristol, England; the Dog Cognition Lab at the University of Western Ontario; and the Canid Behaviour Research Laboratory at Dalhousie University, Halifax.

There seems no end to what dogs might accomplish. Researchers and reporters make pilgrimages to Spartanburg, South Carolina, to meet Chaser, a border collie belonging to retired Wofford College professor John Pilley, who knows 1,022 toys by name (the dog, not the professor). If a *new* item is added to her mountain of toys, balls, and Frisbees, and Chaser is asked to fetch something she's never heard of before, she figures it out in a mental maneuver known as "inferential reasoning." Now Professor Pilley has begun teaching Chaser sentences, like "Take ball to Frisbee."

Questions that would have been too embarrassing for researchers to pose a short time ago are now being asked and investigated. Do dogs think? Do they experience not only primordial emotions and classic emotions but complex emotions like loyalty, jealousy, guilt, and love?

Do dogs possess a rudimentary "theory of mind"—the concept that other individuals have points of view that differ from one's own?

Can dogs read human moods and intentions? Are they happy to be in our company? *Do* they seek help for humans in trouble?

Not all the answers are yes. But all the answers are incredibly interesting.

WHAT SCIENTISTS MOSTLY THINK IS that the epic literature about loyal and heroic dogs has not been put to the test of controlled experiments.

Still, these days as never before, researchers can ask all sorts of questions about the canine state of mind, and then investigate.

"Do Dogs (*Canis familiaris*) Seek Help in an Emergency?" asked two psychology researchers at the University of Western Ontario. "It has long been suggested that untrained dogs are sensitive to human emergencies and may act appropriately to summon help," wrote Krista Macpherson and William A. Roberts. But what if a rescue is coincidental rather than deliberate, inspiring the rescued human to "overenthusiastically interpret and report it? . . . Did the dog *accidentally* do the right thing or did the dog understand the nature of the emergency and intentionally perform life-saving behavior?"

The researchers began by breaking down the classic rescue episode into three components: (a) a dog recognizes a human emergency; (b) the dog approaches a second human to seek help for the first human; and (c) the dog communicates the emergency to the second human.

Lassie: *bark bark bark!*

Mr. Martin: "Lassie, what is it, girl?"

Lassie: *bark bark bark!*

Mr. Martin: "Oh no! Timmy's in the well!"

IN "EXPERIMENT 1: THE HEART ATTACK," a volunteer, walking across a meadow with his or her dog, feigns a heart attack and falls down: "Upon collapsing, the owner remains motionless for six minutes, as the dog is filmed for its reaction to the situation. One or two human bystanders are nearby and could act as a source of potential aid for the victim."

In "Experiment 2: The Falling Bookcase," a dog accompanies its owner into a room. Suddenly a bookcase pitches forward, knocking over the human and pinning him or her to the floor.

"The victim cries out in pain and asks the dog to get help. A bystander whom the dog has already met is available in a nearby room." Again, the dogs are videotaped for six minutes.

The results turned out not to be the stuff of breathless news reports and viral internet videos.

In both experimental groups, dogs approached their fallen owners and stayed close to them. So far, so good. A couple of dogs made some sounds and some nosed the fallen bookcase. But few dogs went anywhere near the bystander and no dog touched the bystander, made eye contact with the bystander, or tried to attract the bystander's attention in any way to the ill or injured owner.

In answer to their question, "Do dogs recognize an emergency situation as such and intentionally take action to help a victim?" Macpherson and Roberts concluded: "The fact that no dog solicited help from a bystander—neither when its owner had a 'heart attack' nor when its owner was toppled by a bookcase and called for help—suggests that dogs did not recognize these situations as emergencies and/or did not understand the need to obtain help from a bystander . . . It is concluded that dogs did not understand the nature of the emergency or the need to obtain help."

In defense of dogs, one must note that while testing dogs on whether they understood emergencies, the research group presented artificial emergencies. The dogs whose owners suddenly lay down in the middle of a field might have known perfectly well that their owners weren't distressed, sick, unhappy, or in pain; perhaps the dogs believed their owners had been felled by a sudden urge for a nap. The researchers acknowledge this design flaw, while explaining their inability to stage experiments in which volunteers actually suffer heart attacks or are crushed by falling furniture.

Thus a scientific confirmation of the Classic Dog Rescue Scenario remains elusive, while anecdotal evidence continues to accumulate.

In full disclosure, I've turned to my own dogs for help now and then over the years. One summer night, Theo, an irascible and headstrong miniature wire-haired dachshund with boisterous eyebrows, shot off like a bullet during a thunderstorm and didn't return. After the rain let up, our family called and searched and splashed through puddles. Lily, then eleven years old, our second daughter and fourth child, could barely sleep or eat during Theo's absence. "Where *is* he?" she kept asking everybody. "But where do you think he *is*?"

"THEO!!!" she cried into the night. We hiked in the woods and we walked around the neighborhood, tacking up flyers. Naturally we turned for help to Theo's longtime companion, Franny, a plump, freckled, well-intentioned rat terrier.

"Franny, can you find Theo?" we asked. From her dog bed, she looked over at us and assumed an anxious expression, but she didn't get up. "Franny, Theo is lost. Where is Theo?" Her creased forehead seemed to relay that she shared our concern, but—when we all headed outside for another search—she pranced alongside us rather than nosing ahead through the woods to pick up Theo's scent.

"Franny," I said. "Theo is your best friend. You spend all day every day together. Until last Wednesday. Can you try to remember Theo?" Happily she hopped through some high grass like a jackrabbit. "Franny, your sense of smell is ten thousand to one hundred thousand times keener than ours. You can smell one part per trillion," I reminded her. "Now can you *please find Theo*?" Back at home without Theo, she bounced up onto the sofa and circled around for a nap.

A few days later, a neighbor called to say he'd seen a weasel in the pond in his backyard. The sleek black animal was clinging to a rock on the edge. The neighbor looked out again the next day and the weasel was still there, in the same position. On the third day, he and his wife went out to investigate whether the unmoving animal was actually alive. They found themselves peering deeply into the eyes of something that was not a weasel, and fished up a drenched and very unhappy wire-haired miniature dachshund. Lily tenderly carried a soggy and relieved-looking Theo home in her arms and devoted her evening to cleaning, brushing, feeding, and cooing to him.

The people with the pond live three houses down from us. Franny and I had explored within twenty-five yards of that pond several times. For all I know, Franny and Theo made eye contact. Perhaps Theo relayed, *Can you get me the hell out of this freaking pond?* and Franny relayed, *I'm busy, sorry, though I don't know why on earth we keep walking in circles calling your name instead of fishing you out of the pond. Who knows what goes through these people's minds?*

We did not call the local news media with a heroic report.

DID BEN SAVE KAREN'S LIFE? It seems a bit of a stretch to imagine that, by barking, he intended to communicate his owner's emergency to a second human on the other end of the phone line. Maybe he barked to tell Karen to wake up and answer the phone, or maybe he urgently needed to go outside and use the bathroom. But his behavior that night *was* an innovation. "Ben was not trained to knock the phone off the hook and bark," Karen said. "I think somehow he thought it would bring help if he barked long enough." And it did.

Karen recovered. She grew stronger. At work managing the adult day care center, she motored busily about in her electric wheelchair with Ben at her side; occasionally she stood and even walked a few steps; after work she took Ben to parks and woods to run and to play with other dogs. While he tore about, she chatted with other dog owners. She took up therapeutic horseback riding and made friends there. As she zipped around the grounds of the day care center or across the dusty yard of the riding stable, the wind brushed the bangs from her face, the sun pinked up her cheeks, and she looked young again. She felt young again. Ben, brilliantly black in bright sunlight, a streak of speed and joy across the landscape, always flying to her side the instant she called, seemed almost magical to her. Karen, several of whose great-grandparents were Cherokee, called Ben her spirit guide.

FEELING BETTER, STRONGER, HAPPIER EVERY DAY, Karen began to wonder how many other people might have gotten stranded as she had. How many people had been denied mobility dogs because the service dog agencies had declared them *too* disabled? Were there others lying helplessly in motorized beds in convalescent homes, wasting away?

Then she thought: *What if I start my own agency? I could train four or five dogs a year, as a small nonprofit, for people rejected by the big agencies.*

It felt right and Karen didn't generally engage in much second-guessing. The moment the idea struck her, she knew it was what she would do next, and that she would start immediately. She would invent a new kind of service dog academy. She would set a standard for compassion and generosity toward potential clients finer than anything she'd been shown when at her weakest. She would find and train service dogs for people in the depths of incapacity, sickness, and suicidal despair—her own state, before Ben had rescued her.

Because of course Ben saved Karen's life.

CHAPTER 10

Iyal

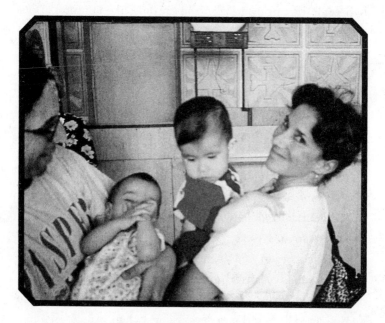

In May 1999, Donnie Winokur, forty-three, of Atlanta, a producer of educational CDs, and her husband, Rabbi Harvey Winokur, forty-nine, beheld the son of their dreams, the child that infertility had denied them.

Twelve-month-old Andrey appeared in a brief videotape recorded in a Siberian orphanage. The pale one-year-old was raised briefly to a standing position by a uniformed caregiver before dropping back to all fours and crawling away. If the couple liked the little boy, they could begin the legal proceedings to adopt him. They liked the little boy very much. By the end of the tape, they were in love with him. Rabbi Winokur recited the *Shehekianu* prayer, thanking God for allowing the two of them to reach this special season, this beginning of a new chapter in their lives.

"Rewatching the footage now, I see that the baby was hypotonic—floppy from lack of muscle tone, like a rag doll," Donnie told me, "but, at the time, I had no idea such a term even existed." She's pretty and trim, pert and intelligent, with dark cropped cowlicked hair and a wry look—a pursed-lips lemony expression softened by compassion. We were seated in the family's high-ceilinged kitchen in an upscale cul-de-sac in a North Atlanta suburb. Sliding glass doors opened onto a deck filled with flowers and hummingbird feeders. "I had done a lot of research on the problems children from Russian orphanages might experience. None of it seemed to apply to this little boy. All I could see was an adorable toddler with thin hair, translucent-looking skin, and deep brown eyes."

They said yes to Andrey and renamed him Iyal (pronounced EYE-al), which means "courage" in Hebrew. Feeling this might be their only chance to become parents, they also said yes to an unrelated little girl two days younger than Iyal, housed in a different orphanage. They renamed her Morasha. Four months later, the Winokurs went to Russia to claim the children and bring them home. They appear in photos together: the broadly grinning, almost disbelieving parents experiencing the happiest days of their lives; and the two unacquainted children passive in the arms of the strangers cradling and kissing them.

They arrived in Atlanta on August 28, 1999, amid a torrent of congratulations, baby showers, gifts, cakes, helium balloons, and a big congregational welcome. (Harvey Winokur is the founding rabbi of Temple Kehillat Chaim, a Reform synagogue in suburban Roswell, Georgia.)

Morasha—dark-haired and petite like her new mother—bloomed into a bright sparkly little girl, but Iyal was irritable, often distressed. At night, he had trouble falling asleep and he awoke frequently, screaming from nightmares. During the day, he seemed hyperactive, unable to calm down or focus; he was also clumsy. If Donnie shyly mentioned to a friend or temple congregant that she really had her hands full with Iyal, the well-wisher might reply: "He's all boy!" or "Mine was the same way!" So she and Harvey felt hopeful, believing they only needed to survive a rocky transition period. For the first eighteen months in Atlanta, Iyal's difficulties floated lightly on the surface of his parents' deep bliss: "God gave us children!"

BUT IYAL BECAME MORE OPPOSITIONAL and explosive rather than less. He exhausted his parents, and scared them at times. "Some time after the children's third birthdays, our wonderful fairy tale of adopting two Russian babies started to break apart," Donnie said. At three and a half, Iyal was a sturdy, big-hearted boy with a wide and open face, shiny black hair in a bowl cut, and a cute hiccupping giggle. But, triggered by the sight of a sticker on a banana, a cartoon image decorating a plastic cup, or one of his sister's bikini-clad Barbie dolls in the bottom of the empty bathtub, Iyal threw large-bore tantrums that shook the house for hours. One day it was the sight of a paper napkin that had fallen to the floor of the car that unhinged him. He unbuckled his seat belt and tried to open the back door of the moving car. He stuffed himself at mealtimes with an inexplicable urgency. He exploded in fury if his mother touched his food, claiming he could smell her on the food. "Certain pieces of my jewelry set him off and I had to stop wearing them," Donnie said. He was hyper-verbal, a machine gun of repetitive, nonsense questions, babbling pointlessly in a symptom the Winokurs later learned was called "forced speech." At preschool, Iyal plowed his tricycle into other children without remorse, or maybe without aware-ness. He ran up to strangers to kiss them, or fell to the ground to feel their toes. The Winokurs' friends and congregants began to fall silent, out of shared concern.

Every night he awoke, got out of bed, and stormed through the house toward his parents' bedroom. Donnie said: "I abso-lutely panicked when I heard him coming. He was a tornado. He was twisting my world inside out during the day and destroy-ing my rest at night. I assumed everything was my fault, that I wasn't a good enough mother."

"You couldn't help but compare the two children," Harvey told me. With an expression inclined toward the commiserative, the rabbi wears a carefully trimmed brown beard and close-fitting wire-rim glasses. He has clearly been gratified by the successes in his life, while humbled by the chaos introduced by his son. He had perhaps anticipated a life less filled with drama. "It was obvious that Iyal's cognitive and emotional development was nothing like Morasha's. Iyal's disabilities began to define our family's existence."

For more than a year, child psychiatrists, pediatricians, and specialists examined Iyal without reaching consensus. The Winokurs racked up what other besieged parents have called the "alphabet soup" of diagnoses: oppositional defiant disorder (ODD), sensory integration dysfunction (SID), attention deficit hyperactivity disorder (ADHD), autism spectrum disorder (ASD), and reactive attachment disorder (RAD), any one of which could rewrite a family's trajectory for many years, if not permanently.

And yet . . . somehow the alphabet soup didn't capture everything. Something deeper rumbled beneath the surface. When Iyal was four years, three months old, he was seen by a developmental pediatrician, Dr. Alan G. Weintraub. This doctor noted microcephaly (small head), short palpebral fissures (small eye openings), epicanthal folds (extra skin folds close to the nose), and midfacial hypoplasia (the middle area of his face appeared flattened). He had unusual creases on his palms and earlobes. When he became anxious during the exam, he made animal noises and tried to escape. Asked to interpret benign pictures, Iyal detected scary themes. The doctor's conclusion was a blow the Winokurs had not seen coming: their son's brain and central nervous system had been severely, irreversibly damaged in utero by alcohol. Though alcohol consumption by Iyal's birthmother could not be documented, the available evidence clearly pointed to fetal alcohol syndrome (FAS), the most extensive form of the range of effects known as fetal alcohol spectrum disorders (FASD).

According to the National Academy of Sciences: "Of all the substances of abuse (including cocaine, heroin, and marijuana), alcohol produces by far the most serious neurobehavioral effects in the fetus." The American Academy of Pediatrics warns: "Even one drink [during pregnancy] risks the health of an unborn baby." The National Organization on Fetal Alcohol Syndrome (NOFAS) states simply: "No safe time. No safe amount. No safe alcohol. Period." These were not messages that Iyal's Russian birthmother had received.

Iyal would never lead a normal life. He was intellectually impaired, with borderline cognitive functioning. The frontal lobe damage in his brain weakened his "executive function," his ability to grasp cause-and-effect, to learn from experience, and to reason. He would be at high risk for a range of secondary disabilities, including impulsive and dangerous behaviors, low social intelligence, social isolation, school failure, unemployment, drug and alcohol abuse, imprisonment, and mental health problems, including suicidal ideation, inability to live independently, and inappropriate sexual behavior. Few medications, therapies, or teaching strategies could be recommended to Iyal's parents as truly effective.

THE SADDEST OF ALL ANIMAL PHOTOGRAPHS and videos are those of the victims of Harry Harlow's "social isolation" experiments. In the 1950s and '60s, the American psychologist raised baby rhesus monkeys alone in dark cages, with nothing but wire structures for company. He called the stainless steel enclosures the "Pit of Despair." In experiments that would be deemed unethical today, he produced profound depression in primate infants. The sorrow and terror of each tiny motherless animal—alone in the universe—pierce the heart. You can see "Number 106" in a video, a wide-eyed baby sucking from a bottle attached to the "wire mother," who then flees to the "cloth mother," a wire figure with a robot-like clown face wearing an apron of carpet fibers. This is an Unlikely Friendship of the loneliest and most unnatural kind. Robbed of love and mothering, the baby monkey grows into a dysfunctional, self-stimulating, self-mutilating adult who— when finally inserted into a troop of other monkeys—doesn't know how to act and remains a social isolate forever, shunned and bullied by the others.

The world didn't need Harlow's pitiless experiments to demonstrate that solitary confinement is the cruelest of all punishments for a living thing, especially for a juvenile. Any animal cut off from its kind as a result of a natural disaster or captivity will, as in the stories of interspecies friendships, reach out to someone, anyone, nearby; or, in the case of the baby monkeys, to a softer inanimate object. It is an infinitely cold universe for any creature who finds itself alone, and for young creatures most of all.

Trolling the internet for a miracle, Donnie discovered a dog academy in rural Ohio training assistance dogs for children with autism. Could a service dog help connect Iyal to the world?

"SHE HAS NO FRIENDS. NOT ONE. She has never had a friend."

"We say he's in his own zip code. No one else lives there."

"I think of her as alone in the universe."

The parents applying to 4 Paws have described their children in these terms. In the photos and videos submitted by the families, some of the children display the wide-eyed look of stunned incomprehension and banishment of Harlow's baby monkeys. There aren't a lot of happy age-appropriate social options for school-age children and teens who require strollers, spoon-feeding, diapers, breathing tubes, feeding tubes, or padded helmets. Their parents may fill the children's hours with therapeutic equipment, tutors, drivers, therapists, psychiatrists, and special schools, while what a child wants and needs most desperately is a friend. Often a child longs for a friend despite rejecting everyone who reaches out. Such a child may be convinced to come out of hiding if the proffered playmate is a dog.

Dogs seem to radiate such empathy that even the most challenged and withdrawn child is encouraged to give friendship a try. *Is* empathy what we're seeing in dogs? Or is it possible that—as in the case of "Do Dogs Seek Help in an Emergency?"—dogs sometimes, *accidentally*, display emotionally appropriate behavior and then find themselves lavishly praised and rewarded for it?

Empathy is a complex emotion just at or beyond the frontier of our current understanding of the inner lives of animals. It is a current subject of investigation.

DOG LOVERS, ACCORDING TO POLLING DATA, behold many complex emotions in their dogs (probably far more than non-dog-lovers can detect in the animals). They perceive Trust (rolling on the back, showing the tummy, hoping for a tickle), Hope (eyes wide, lips parted just shy of a smile, *I'll have what she's having, but make mine medium-rare*), Suspicion (lowered eyebrows, eyes darting beneath partially-lowered lids, *I believe I just heard you unwrap a sandwich. What have you done with it?*), Gratitude (that humble look on the face, ears flat and eyes softly gleaming, when the dog is offered the last few licks of your ice cream cone), Boredom and Loneliness (what else would inspire an individual to chew the foot-rest off a Laz-E-Boy, scatter shreds of a 12-pack of toilet paper across the house, or eat a bottle of glitter?), Nervousness and Anxiety (*Thunderstorm! Balloons! Fourth of July! The vet's office! OMG!*).

Pride was perceived by 65 percent of dog-owners in a British poll. Dr. Alexandra Horowitz agrees: "You can tell a lot about a dog by observing how he carries his head," she writes. "Think of a dog prancing around in front of other dogs, tail and head high, with a cherished or stolen toy . . . this is a clear, intentional gesture of something like pride."

Loathing and Contempt don't make the list of what dog-lovers see in their friends, while Shame and Guilt have gained thousands of proponents through internet memes of dogs looking sheepish when they're caught doing something ridiculous. A video of a dog confusedly wearing a swinging trash can lid around his neck (while standing in a kitchen strewn with garbage) has attracted half a million viewers. But some researchers believe that what provokes the "guilty look" is less the dog's misbehavior than a human's *assessment* that the dog has misbehaved. "Many have claimed . . . that dogs are capable of feeling ashamed when they violate some rule, for example when they chew the master's slippers to bits and the crime is discovered," writes Dr. Csányi. "The problem with this phenomenon is that when the corpus delicti is discovered, the master usually turns to the dog with a portentous tone of voice: 'What have you done?' to which the dog responds with the usual behavioral forms of the subordinate. My inclination is to attribute this more to the fact that the question always contains an element of a threat, and dogs probably recognize this and fear the reproach . . . Shame is an emotion of very high order and, among humans, it is the expression of very complex social relations, which, I believe, dogs did not need during the course of domestication."

Jealousy is recognized by 80 percent of polled dog lovers. I know it well. My dog Zizou doesn't approve of my visiting my daughter's house, where dwells Zizou's arch-nemesis, the slender, guileless, bewhiskered Ivy. When I come home from an Ivy encounter, it's no use pretending otherwise. No matter how great a show I make of grocery bags, as if implying, "I just went shopping!" Zizou rams my legs again and again hard with her muzzle. *I . . . cannot . . . even . . . believe . . . what . . . I'm . . . smelling,* she seems to express, one hard knock at a time. Karen Shirk has seen it, too, in Abel, the tricolor papillon vying against Piper for Karen's attention.

Dr. Christine Harris, a psychologist at U.C. San Diego, tested for Jealousy in dogs. She modified an experiment that had been run elsewhere with human infants, who were observed while their mothers focused on either a realistic-looking doll or a book. In that study, babies exhibited "behaviors indicative of jealousy . . . when their mothers interacted with what appeared to be another infant (but was actually a realistic-looking doll). The infants did not display the same behaviors when their mothers attend[ed] to a nonsocial item (a book)."

Dr. Harris invited dog owners to pet and talk to plush toy dogs while their own dogs watched, then to pet and talk to a jack-o'-lantern, and then to read a children's book aloud. The dogs paid little attention to the jack-o'-lanterns and even less to the books, but "when dog owners petted and talked to a realistic stuffed dog that barked and whined, the people's own dogs came over, pushed the person or the stuffed dog, and sometimes barked. After the experiment, many of the dogs sniffed the rear end of the stuffed dog, suggesting . . . that the dogs thought it might be real." Dr. Harris concluded that the dogs "showed a 'primordial' form of jealousy, not as complex as the human emotion, but similar in that there is a social triangle and the dog is trying to make sure it, not the rival, receives the attention."

Not all scientists are convinced that this was a proof of Jealousy. "Some argue that jealousy requires complex thinking about self and others, which seems beyond dogs' abilities," writes James Gorman for *The New York Times*. They feel that, while there may be something going on here, it may not be Jealousy.

Nevertheless, it might be wise for Piper to watch her back.

DONNIE WINOKUR HAD NO IDEA of the debates being held in scientific circles over whether nonhuman animals possessed the capacity for complex emotions and, if so, which animals and which emotions. What she had learned from researching the experiences of other parents with special needs children was that kids calmed down, looked happy, and communicated when in the company of their 4 Paws for Ability dogs. Dogs had inspired children to laugh, to make eye contact, to venture out into public, and even to speak for the first time.

"Human beings are endlessly interested in animals," Dr. Beck told me. "A child is *born* interested in animals, hardwired to want to look at animals. As the child grows, a dog, in very real ways, becomes a surrogate sibling, naturally playing, sharing, competing, and bonding. Animal friends also offer children vital chances to nurture, a skill-set especially valuable for little boys, who have fewer opportunities in our culture to express caregiving behaviors.

"When you're a kid, a dog is a natural best friend," he says. "A dog seems to understand what you're saying, while giving no negative feedback. The animal is always there, genuinely interested in what you're doing, with the added benefit of relaxed, close, physical contact. If you're saying something disgusting, the dog is still interested. The animal will never betray you because the dog is not going to relay details to your parents. Far better than a doll or a stuffed animal, an animal responds to you. The dog becomes your first true confidante and best friend."

In 1887, the Reverend Henry Ward Beecher expressed something similar from his pulpit in Plymouth: "The dog was created especially for children. He is the god of Frolic."

WHEN DONNIE PRESENTED the fruits of her research to her husband, he recoiled physically. "Now we need a *dog*?! Are you *kidding* me? We don't need a dog!" cried the rabbi, certain that one more howl raised under this roof, one more living creature whining for attention, one more source of strife between the warring children, and one more obligation standing between him and complete collapse into unconsciousness at bedtime would push him beyond endurance. "No, Donnie. It's too much. I couldn't take it. *No,* absolutely not."

"Harvey, this could really be the help we need," Donnie wheedled.

"Forget about it. *Please,*" he groaned. "It's me or a dog."

CHAPTER 11

Shelter Dogs

A German shepherd mix slated for euthanasia watched Karen from behind the bars of his cinder-block cell in the cacophonous county animal control building. With his long black muzzle and imploring brown eyes, he looked at her with that heartbreaking shelter-dog mix of worry, fear, confusion, and hope. "This is a good-looking boy. Do you know anything about him?" Karen called from her wheelchair to a nearby worker. "Can he sit? Can you sit, boy? *Sit.*"

The dog sat. His haunches trembled with the sincerity of his "Sit." He tentatively raised one paw a few inches above the floor, in case the stranger also wanted "Shake." She didn't say "Shake," so he lowered his paw quietly and put his whole focus back into his excellent "Sit."

He was an "owner-surrender," though there was no coercion or "surrendering" about it: his people, for reasons unknown to the shelter, had brought him here to be disposed of. In crowded shelters, owner-surrenders are among the first to go: without the required ten-day "stray hold" bestowed upon lost dogs or cats for whom someone may be searching, the owner-surrenders quickly join the ranks of the sick, the injured, the elderly, the pregnant, the nursing mothers and their newborn litters, and the defamed pit bull breeds—no matter how gentle—to be euthanized one by one by one, usually by lethal injection.

WORD HAD SPREAD QUICKLY AFTER Karen shared her great idea with a few friends and colleagues. She was right about the clean power of the idea; it seemed to have a strong spring wind behind it already. Within a week, a friend with paraplegia asked if Karen would train a mobility assistance dog for her. A professor of Middle Eastern history at the University of Cincinnati, weakened by fibromyalgia, asked for one, too. Then the mother of a local twelve-year-old girl phoned Karen from a Cincinnati hospital: her daughter had been stricken with a rare spinal stroke and now lay paralyzed from the waist down. Every service-dog agency had denied their application because mobility dogs couldn't be placed with children. Would Karen please give them a dog?

Why won't the agencies place dogs with children? Karen wondered, and asked for a few days to do some research. At the library, she read the Americans with Disabilities Act and saw no legal impediment to placing a dog with a child if a parent served as co-handler. Unaware that she was crossing the Rubicon among service dog agencies, she called back the family and said yes, she would train a dog for the child. She would soon be met with hostility and derision as industry people learned that she'd flouted their unspoken rules.

In October 1998, a group gathered at Karen's cabin to become the board members of the nonprofit corporation, 4 Paws for Ability. The following month she flew with Ben to a conference of service dog agency directors in Seattle to introduce her agency as the new kid on the block. "I rolled into the reception and announced, 'I want to place dogs with children,' " she says. "They all said, 'You can't do that!' And I'm like, 'Hello, I'm going to do it.' They tried to give me reasons why not, like: 'If you give a dog to a child, then when the child is a teenager the dog will grow old and die and the child is already disabled and going through enough without having his dog die.' I'm like, '*What?* That's not a reason.' Then they said, 'Well, a disabled kid can't handle a dog.' So I said, 'The parents can,' and that was the last straw with them: 'Oh no! That's not allowed, that's not legal.' And I said, 'Where does it say that? Show me the law.' And they couldn't show me, because it's nowhere in the law that parents can't be the handlers for their child's dog."

Back in Xenia, Karen advertised for and hired local dog trainers. Then she made the rounds of animal shelters to find a few good dogs.

BEYOND THE HAPPY REALITY OF perhaps 150 million doted-upon pets in the United States lives a shadow population of caged, doomed animals. Close to 8 million companion animals enter animal control units every year: they've been picked up as strays, surrendered by their owners, rescued from puppy mills or animal hoarders or dog-fighting rings, or confiscated in animal neglect or cruelty cases or during domestic or child abuse arrests. About a third of the dogs and 40 percent of the cats will be euthanized. In many animal control units, one day each week is referred to as "PTS Day." Put To Sleep Day.

"People will say, 'My kid promised to take care of the dog and he's not taking care of it,' " Jamie Martinez, of DeKalb County Animal Control in Georgia, told me. "And I'm like, 'How old is your kid?' and they'll say, 'Eight.' Or people don't do their research. They get a border collie puppy with no idea how to handle a high-energy herding dog. They buy a shih tzu without understanding the level of grooming required—then they drop him off here so covered with mats and knots he can barely see or move. Sometimes families are losing their homes and have to give up the pets; those are just really sad, the kids are crying, the parents are crying. Other times people just didn't think it through. They say, 'I work all day and no one pays attention to the dog,' and I wonder, *What did you think was going to happen?* Tons of people say, 'Oh, I'm moving and I can't take the dog,' and I'm like, 'Could you pick a different apartment?' Sometimes the people tell their pet goodbye, but usually they just turn around and leave. A dog who's coming in as a stray, or who has spent a long time chained up in a backyard, seems relieved at first to find food and shelter here. But a house-pet just looks really sad and really confused."

THE SCRAPE OF SHOVELS AND SPLASH OF WATER and the homesick yelps of imprisoned dogs ricocheted around Karen and the German shepherd mix as the dog sat for her on the cement, making worried eye contact, in the most important and possibly last audition of his life.

Did the shelter dog understand on any level that he had captured Karen's attention, however briefly?

To pose the question scientifically: Is one animal aware of the cognitive focus, the "attentional state," of another animal?

I recently, accidentally, conducted an experiment on "attentional states" with my dog Zizou. She is a white dog with black spots, brown cheeks and eyebrows, a mosquito-thin waist, ears that flip over at the tip, a long muzzle, and a fishhook tail. She looks like a rat terrier and we'd wanted a rat terrier and she was adopted from a rat terrier rescue organization; but her Wisdom Panel results detected no drop of rat terrier in her. No ancestor of Zizou's ever got within 100 yards of a rat terrier. I sat down on the sofa in the family room with a dish of ice cream and attempted to enjoy it despite the unblinking gaze of four dogs—one on either side of me on the sofa like bookends, facing in; the dachshund on the ottoman, legs extended in front of him, facing forward; and Zizou on the floor across the room. I put up my knees, with a blanket over them, to shield the bowl of ice cream from everyone's view. I *was* planning to share, but not *yet*. Zizou wasn't looking at the ice cream, though; she was looking at my face. She was gazing relentlessly at my eyes. Whenever we made eye contact, she relayed with hypnotic intensity: *Give me some of your ice cream.* To avoid her, I leaned my knees a fraction of an inch to the right, blocking Zizou's eyes from mine. I turned to my ice cream. When I glanced up, I saw that Zizou had corrected for my knee interference and had moved ever-so-slightly to her left, on the far side of the room, so that our eyes connected again and a frisson of her desire passed between us. Deciding to test her, I slid down the tiniest bit—you really couldn't tell I had moved at all—except that the beam from her eyes into mine was broken again by my knees. Zizou didn't appear to move a muscle either, but then—like sunrise above a mountain range—her eyes, round and shining, very slowly rose above my blanketed knees. *Give me some of your ice cream.* I don't know if this animal possesses theory of mind or not, but she definitely knows whether or not she has captured my attention, after which she captured my ice cream.

My daughter Molly's dog, Ivy—a pale-yellow, lean, whiskery girl with light brown eyes—enjoys an evening game of "Raccoon." She initiates it by grabbing her almost-life-size plush toy, shaking it, then dropping it in Molly's lap. Molly hurls Raccoon across the room. Ivy races for it, "kills" it again, and drags it back to Molly. Ivy could go on like this all night, but *only* when she has Molly's full attention. The moment Molly glances down to check her phone or turn the page of the newspaper, Ivy stops what she's doing and leaves the room, in something of a mood. "Sometimes I can coax her back if I retrieve the raccoon and shake it myself," Molly says. "But sometimes not. Isn't it such a human response? *Fine, if you've got better things to do than spend time with me, I'm outta here.*"

Two family dog anecdotes do not a scientific theorem make, but the question of whether dogs are aware of "the attentional state of others" has been scientifically probed, as well. Dr. Alexandra Horowitz calls this "attention to attention" and dogs appear to be hardwired for it. Before issuing an invitation to play, for example (like with a "play bow"), a dog will confirm that his potential playmate is in a position to *receive* his invitation. If not, he will—with a bark or a bump or a nip—catch the other dog's attention first and then execute a "play signal."

Researchers with the Family Dog Project at Eötvös Loránd University wondered whether dogs are aware of *human* attention. They designed numerous experiments in which the subject dogs could discriminate between an "attentive" human and an "inattentive" human. In one scenario, a dog was led into a room, at the far end of which sat two humans eating sandwiches. One person wore a blindfold; the other did not. The dog was released and permitted to approach and to beg (stare at the person, stare at the sandwich, etc.) from the human of its choice. Not very surprisingly, the dogs significantly preferred to approach humans not wearing blindfolds.

(Zizou and I could have told them this and saved them some time, but scientists prefer to do their own fact-finding.)

Was the shepherd mix who sat so politely in his cage for Karen, unflinchingly and longingly gazing into her eyes, *aware* that he'd captured the attention of a human being, something in scarce supply in a county animal shelter? Of course he knew. He was begging her, with his eyes, not to leave him.

"I'M GOING TO GIVE HIM A TRY," Karen said to an employee. "Let's take him outside." The worker stepped into the pen and clipped a leash to the dog's collar.

On the way down the cement hall toward the steel exit door, the shepherd, leashed, stayed beside Karen's wheelchair, but his paws moved double-time, like a speeding cartoon character whose legs accelerate into wheeling blurs. Outside, the dog blinked in the sunlight and barely knew which way to run first.

Just in case, he briefly sat again, tremblingly, joyfully. When the passenger door of the van opened to him, he bounded into the seat, wiggled in happiness, settled in, and never looked back. He moved into the cabin with Karen and Ben and soon began training for the twelve-year-old girl with paralysis.

Soon two rescued golden retrievers joined them, one for each of the adult women who'd requested dogs. It was a happy messy life for Karen, the start of her finding a way toward the life she wanted. The hospitalized preteen squealed with joy when she saw the German shepherd mix for the first time and named him Butler—"because he's going to be like my furry butler!" When his mobility training was finished and he was placed with the family at home, Butler broke the no-child barrier among service dog agencies. He was a great success! He heeled beside her wheelchair, slept on her bed, and always sat up extra straight and tall when told to sit, since this was evidently his winning skill. The girl's laughter rang through the house again whenever Butler, unable to contain his love and happiness, stood up, propped his front feet on the armrest, and leaned into the wheelchair to lick her cheeks.

"AM I TOO OLD FOR ONE OF YOUR DOGS?" strangers phoned to ask Karen. "Is my child too young for one of your dogs?" "Am I too disabled?" "Am I disabled enough?"

Karen told everyone the same thing: "If your life can be improved by a dog and you can take good care of a dog, I'm going to give you a dog."

A couple with a ten-year-old son with autism phoned to say that their boy constantly ran away and they'd hoped a service dog might keep track of him, but the service dog agencies had all denied them. This was again new territory. Karen knew that placing service dogs with adults with invisible disabilities, like post-traumatic stress disorder or seizure disorder, was the cutting edge of service dog work, but it hadn't yet been tried with children. It was a tall order, quite different from training Butler for mobility work with a child.

Back to an animal shelter she went.

Despite the forbidding prison-like appearance of the place and the collective hysteria of the stressed and frightened dogs, Karen knew there had to be animals there with high intelligence and fine dispositions. The problem was that their panic at the harsh, crammed-in, and grating conditions of captivity concealed their true natures. The confinement in cement cells with industrial drains in the floor made the dogs seem ferocious, impossible to tame, even insane. They bared their gums and barked in fear, scaring away adopters.

As Karen wheeled through the cat room on the way to the dog kennels, cats stuck their forearms through the bars of their stacked-up cages, waving their paws around in blind search for human contact. Karen stopped to stroke the arm of one cat; the lean middle-aged tabby instantly withdrew his arm and flipped onto his side in the cage in winsome appeal. He'd waited so long for a tummy-rub! He stretched out and began to purr. But Karen couldn't reach that far into the cage and had to move on. She knew that virtually none of these adult cats would see daylight again.

TAIL LOWERED, EARS FLATTENED, FACE DOWNCAST, Patches, a beagle mix, managed just a couple of tentative halfhearted tail-wags from the back of his cell. His overtures hadn't beguiled anyone in the nearly twenty-one days of his captivity and his time was up. Karen positioned her wheelchair outside his cage for a closer look. Every morsel of emotion rushed into the dog's moist trembling nose. He approached and shyly pushed his nose through the chain-link barrier. "Okay, boy, I see you," she said.

When he was led out of his cage by a handler for one-on-one time with Karen, the little dog was so excited, shaking so hard, he couldn't avoid peeing a little on the cement floor. Like Butler before him, he left the shelter riding high in the passenger seat of Karen's van, his mouth wide open with happiness, his ears rippling in the wind he hadn't felt in a long time.

Before pulling onto the state road, however, Karen sighed, stopped, wheeled around, pulled back into the parking lot, and called out her window to a staffer to bring her the middle-aged tabby cat.

PATCHES, THE RESCUED BEAGLE MIX, became one of the first dogs in the world (similar work was beginning in Canada at that time) trained in autism assistance. He may have become the first dog in the world trained to track a single child. Now when their son disappeared, his parents cried: "Patches! Find Kevin!" And Patches took off to find the boy, wherever he was. One night he tracked him to a stranger's backyard three blocks away. The land sloped down to a stream; Kevin, in his pajamas, was peering into the water when the dog interrupted his reverie. "Patches just saved our son's life again," the parents emailed Karen.

The cabin filled up with rescued dogs. "It's a wonderful feeling when we see one of our animals adopted by 4 Paws!" said Mary Lee Schwartz, executive director of the Humane Association of Warren, Ohio. "We're happy when a dog gets adopted to a normal home, but when one gets adopted to a home where he's going to help someone, we're thrilled! I can't think of a more exciting thing to happen for a dog, especially one on Death Row." Another shelter worker commented: "People are surprised that we have such highly talented dogs coming through our shelter, capable of performing the functions of service animals. But of course we do."

All shelters have them: indescribably marvelous animals just waiting to be given a chance.

"I'D BEEN THINKING SMALL," Karen says now. "I'd been thinking five, six dogs a year. By around this time, I realized this was not going to be small. I couldn't have a day job and keep 4 Paws going on the side."

In 2000, Karen resigned from her job managing the adult day care, rented a house in town, and turned full-time to 4 Paws. Jeremy Dulebohn—the trainer from Ben's puppyhood—answered one of her Help Wanted ads and began training dogs for 4 Paws. Soon 4 Paws was his biggest client, then 4 Paws was his only client, and in 2001 he joined the agency full-time as training director.

"At the start, we trained a little bit of everything, most of them rescues," Jeremy told me over lunch at Applebee's. He and Karen recalled a boxer, a rat terrier, beagles, bloodhound mixes, an Australian shepherd, a collie, a Chinook, a Siberian husky, a Bernese mountain dog, and every sort of mix.

"The rat terrier was interesting," Karen said. "He did fine, until you dropped a piece of paper. Then he went insane and ripped it to shreds. We couldn't cure him of it. We gave him to a college student for free. We told him, 'Great little dog, but if you're going to drop your term paper near this dog, you better make sure you've got a copy.'

"I told the kid: 'Don't let him see your folding money!' " said Jeremy, his shoulders heaving in laughter.

Service dog work is challenging! One overview lists a few of the distractions that may besiege a working animal: "Cars careen by with . . . music blaring, buses backfire, people whistle and call in an attempt to catch the dog's attention, shopping carts rattle past at close range, loudspeakers blare 'Blue Light Specials,' and children grab or hug the dog without warning . . . It's asking a lot of any animal to ignore such stimulus-rich environments." Karen and Jeremy quickly discovered how easy it is for a wonderful animal to flunk out of service dog school. A fear of elevators or of mannequins, or a high degree of possessiveness regarding a rubber squeaky squirrel, puts a pup on the path to becoming a nice family companion instead of a working dog. (4 Paws maintains a waiting list of families eager to adopt one of their dropouts, as it will be marvelously well trained in *nearly* every situation.)

"Herding dogs do well," Jeremy said. "But terriers are going to dig and go after varmints; they were bred to do it. Sight hounds, like greyhounds, go after prey. They run thirty-five miles an hour; they see a rabbit and they're *gone*. Imagine having a kid in a wheelchair on the other end of *that* leash."

"The problem we ran into with the German shepherds was their devotion to their trainers," said Karen. "They bonded to their trainers and didn't want to move on. We'd introduce them to their families and they'd just lie there with these mournful eyes. We're experimenting with softening that intense GSD nature, but keeping the intelligence, by crossing them with other breeds."

"Small dogs bred for companionship can work," Jeremy said. "Boxers are smart, but they drool a lot, and they're gassy."

"Bloodhounds are phenomenal trackers, but they're huge, they're wrinkly, they drool, and they have skin problems," said Karen. "So we tried breeding them to Labrador retrievers. Their size was brought down a little, they don't drool, and they're dead-on, they don't mess around; they get on the track and they stay on it till they find the kid."

"Border collies are super-smart," Jeremy said. "They're the most intelligent dogs and their work-drive is the highest. Those aspects are terrific for alerting to seizures. A border collie would never miss a seizure in the night, they're so into their work. But they're mostly too intense to be service dogs and they sometimes get nervous in public. We're wondering if crossbreeding them could produce that same work-drive and attention to detail, but bring some common sense into the picture. Labradoodles seem to get the worst of the poodle, while goldendoodles seem to get the best of goldens."

"People call here trying to donate cane corsos and pit bulls to us," Karen said. But pit bulls have been outlawed in some communities (unfairly to the breed). And cane corsos, which are Italian mastiff-type dogs, descend from Roman dogs of war and were bred to be guard dogs and hunters of wild boar and cougars. In capable hands, cane corsos are wonderful pets, but their enormous size and natural suspicion of strangers terrify those unacquainted with the breed. "I like them, they're trainable," said Jeremy. "There's nothing wrong with them other than if you walk into a store with a cane corso, everyone in the store goes screaming out the back.

"Any dog, any breed can become a service dog," Jeremy said, "but, over time, it looked like around seventy percent of Labradors, golden retrievers, goldendoodles, and German shepherds graduated from our program, but only about two percent of other breeds made the cut. If you take in a hundred rescue dogs, twenty will graduate; but of a hundred dogs we breed and raise here, more than ninety will graduate. For special requests—for a local man with Parkinson's, for example, or for hospice patients—

we'll rescue older dogs. And we find good homes for all the dogs trained here, whether or not they graduate."

Shelters and rescue groups across Ohio began offering dogs to 4 Paws. Breeders phoned to donate puppies not quite up to breed standard. Anonymous local citizens trespassed at night and abandoned cardboard boxes full of squirming, squeaky puppies. Even in winter. Sometimes the boxes contained kittens. An early-arriving staffer found an older dog tied up and trembling by the front door; on another morning, the first to work found an abandoned dog crouching inside a too-small cage, tail between her legs, frightened and submissive. One day during work hours someone *threw* a dog over the tall wooden fence at the far side of the upper field and ran away. "We're not a shelter, people don't understand that," said Karen. "We rescue and train shelter dogs, but just because someone is irresponsible and abandons a dog here doesn't mean the dog's going to become a service dog." Soon they were up to their eyeballs in dogs. Karen had dogs lounging across every chair, couch, bed, and rug in her little house, and Jeremy's backyard in Wapakoneta looked like the Westminster show-ring. They accepted as many as they could, found safe homes for the rest, and hired more trainers.

In 2005, 4 Paws for Ability bought the one-story VFW Hall on Dayton Street in Xenia. Indoor and outdoor pens, kennels, play areas, and training areas were installed, while Karen and Jeremy continued to rescue dogs from shelters and accept dogs from breeders and inherit dogs from unwanted midnight visitors. They sent up a call for help. Local families stepped up to foster dogs, as did college students on nearby campuses. Volunteers and high school students began arriving daily to play with dogs and walk dogs. If there were a dog version of the children's picture book *Millions of Cats,* Karen and Jeremy could have joined the very old man and the very old woman. Hundreds of dogs, thousands of dogs, millions and billions and trillions of dogs.

BEN WON MANY HONORS, including the Delta Society Hero Dog Award, Canine Companion for Independence Certificate, a Temperament Certificate from the GSD Club of America, and Canine Good Citizen Award. He marched and Karen rolled in the GSD Parade of Great Dogs. To this day, framed photographs and newspaper clippings of Ben decorate the halls of 4 Paws, and a lighted trophy case celebrates his achievements.

He was only eight years old when he somehow tripped and fell down one day, and had a hard time regaining his footing. Karen could scarcely believe what she was seeing. Over the next couple of weeks, he had more mishaps, more falls, more clumsiness. "When he walked, he dragged his feet and it ripped his toenails open," Karen says. Her worst fear was confirmed when the vet diagnosed Ben with degenerative myelopathy, for which there was no cure.

Now off the ventilator and out of the wheelchair, Karen slowed down her pace to gently accommodate her old friend. Leaving him home alone was out of the question, so she boosted him up into her van every day onto a throne of pillows and prepared comfortable resting spots for him at 4 Paws. She invented small tasks to help him stay busy and feel useful. "I dropped little items near him, so that he could retrieve them for me," she says. "I'd ask him to smell the contents of a box before I opened it." She became Ben's service human.

Ben had boisterously trampled Karen's fortress of solitude as if it were a sand-castle on the beach, allowing colleagues, friends, clients, more dogs, and ultimately her own children to pour into her life. "Ben made everything possible for me," she says. "Ben was my first close adult relationship. I truly needed him. The closeness I shared with him inspired me to adopt my children." She would name her first son after him.

In April 2002, realizing he was nearing the end, she took him on a weeklong vacation. "I wanted Ben to experience his final moments of life with joy," she says. "In a park, he ran to chase and fetch a Frisbee with some teenagers. His beautiful gait was off, he fell down, he dragged his hind legs for a few yards, but he came back with the Frisbee in his mouth and his eyes sparkling with joy and love of life. I took him on a final walk through the forty acres of woods that had been his kingdom. Then I took him to the vet for his last time."

Online she posted, in language familiar to many dog lovers: "My Ben has crossed the rainbow bridge and returned to the place where all dogs run and play." She is far from alone in imagining a dog heaven.

"If there are no dogs in heaven," Will Rogers famously said, "then when I die, I want to go where *they* went."

An oft-told story (a version of which appeared in a 1962 *Twilight Zone* episode) envisions a man and dog wandering a dusty road in search of the entrance to heaven. Finally they come to the glorious golden gates. Just beyond the entrance, the thirsty pair see fountains and pools of clean water. They're so thirsty! But a white-robed gatekeeper politely blocks them. "No dogs allowed," he says to the man, "but *you* can come in."

The man, looking down at his beloved companion, shakes his head no. The two old friends continue to trudge down the dusty road. They come to another, more modest entrance of some kind, more like a paved driveway into a public park. They glimpse dogs and people playing together on green lawns in the distance. "Welcome to heaven!" exclaims a gatekeeper, setting down a water bowl for the dog and handing a tall water glass to the man.

"What?" says the man. "I thought that was heaven down the road."

"Oh, no!" says the heavenly angel. "That's the entrance for people who abandon their dogs. That was the gateway to hell."

"Karen, you are an amazing woman," a client named Roo Niermann posted on the 4 Paws Facebook page. "When you arrive in heaven one day in the far-off future, you are going to be welcomed by a million cheering animals."

Being welcomed by Ben will be enough.

CHAPTER 12

Prison Dogs

Memories By Beverly

Eddie Hill, age twenty-three in 1989, was a mild-mannered, reasonable guy, but he saw himself as a weakling, a pushover. When he looked in the bathroom mirror, a slight fellow with fuzzy reddish hair, a meek smile, and pale worried eyes, he didn't glimpse promise, or intelligence. He saw "pretty damn worthless." He saw "loser." He was a smart kid who ran with the defiant, troublemaker types in high school; after high school in Martins Ferry, Ohio, he served a brief stint in the air force, took an early discharge, and drifted to Columbus, Ohio, without a job or a place to live. He slept on the couch in his older sister's apartment in exchange for babysitting her children while she worked and he knocked around with sketchy lawbreakers at night. He was heading toward a crisis.

In May 1989, his high school ringleader and ex-brother-in-law, Donald "Duke" Palmer (the father of a couple of Eddie's sister's kids), showed up and asked for a favor. Duke, twenty-four, was a six-foot-one-inch 220-pound black-haired army vet with raggedy facial hair. At sixteen, when Duke had started seeing Eddie's sister, he'd towered over fifteen-year-old Eddie and claimed to feel "protective toward the little guy." Eddie felt browbeaten, but had liked the sense of running with Duke's pack. When Eddie's sister got pregnant, she married Duke, both of them teenagers. They soon had two kids and fought a lot. Eddie tried sharing the rent with them in a Florida bungalow until the noise drove him off, at which point he moved into his car. When Duke's wife kicked him out, he showed up outside the passenger-side door with his stuff, ignored Eddie's quiet demurrals, and moved in. The two of them lived in Eddie's car for a while. When Duke finally moved out in fury, the pair fell out of touch until crossing paths in Columbus in the spring of 1989.

Duke had become a swaggering drunk, cocaine addict, and drug dealer. He wore wide oval orange-tinted sunglasses and saw himself as a ladies' man. Fired from a cleaning job at a Motel 6, he lived off drug sales. Twice he stole prescription pills from his mother, who was addicted to them, and tried to overdose. One night he scared some people by holding a loaded pistol to his head, grinning. He was institutionalized twice.

Eddie was at a low point, too, with a crushing post-military depression. Most nights he crashed at his sister's apartment, watched her three kids while she worked, and felt stuck. He knew it was a lousy idea but slid back into his high school ways of hanging out, getting drunk and getting high with Duke Palmer. He laughed hard at Duke's parties as if he liked those people.

In early May, Duke's sister Angel, who had spina bifida, asked Duke to drive her back to Martins Ferry to pick up her SSI disability check. Duke didn't own a car—never had—but knew who did. And Eddie Hill had never been able to say no to him.

On Sunday, May 7, they loaded up Eddie's brown Dodge Charger with Angel's wheelchair, some hard rock cassette tapes (AC/DC, Guns N' Roses), and a bottle of whisky for the two-hour drive. Duke stashed a .22-caliber pistol and ammo in Eddie's glove compartment. *That gun is ridiculous,* Eddie thought. *He thinks he's a big-time coke dealer.* They spent the night at Angel's house in Martins Ferry, an antebellum country town built along a swerve in the Ohio River, a stone's throw from West Virginia.

On Monday, May 8, 1989, they took Angel to cash her check and then hit her up for cash. They dropped her off at home, drove to a state store, and bought a fifth of 100-proof Southern Comfort. Sharing the bottle, with Eddie behind the wheel and Duke giving directions, they went joyriding across their boyhood county, blasting Guns N' Roses, skimming the blacktop, and singing along at the tops of their voices. Duke fired his handgun out the window at trees and fence-posts. Once he asked Eddie to pull over beside a field. He got out, shot at a Coke can and a plastic pumpkin in the distance, and called it target practice.

Eddie later thought: *Two dumb punks up to no good.*

There were places Duke wanted to drive by—there was a guy

he hated, who'd briefly dated his ex-wife after the divorce; he wanted to lay eyes on him or something. Residual cocaine, fresh liquor, the hoarse raging music, and the rusty velocity of the old sedan stoked Duke's righteous indignation, so he shouted directions and Eddie stepped on the gas. They careened down the two-lane blacktop past cornfields and pastures, squealed around a blind curve, and drove straight into the back of a white pickup truck idling on the road.

SHOCKED, EDDIE TURNED OFF THE CAR and hurried to apologize. The driver, on his way home from a birthday errand for his son, barged out of his truck. Aggravated, he strode toward Eddie, as Duke, falling-down drunk, got out of the passenger seat of the Charger and staggered toward them. The local man "was ranting and raving," Duke would later testify. "He was throwing a fit, like he wasn't even hearing Eddie at all. Eddie kept telling the guy he was sorry, but [the man] was upset and grabbed ahold of, went to grab ahold of Eddie. When he reached for Eddie, I grabbed his arm and swung to hit him."

Coming up behind the stranger, Duke grabbed the man's left arm with his left hand and brought his right hand down to land a blow on the man's head to make him shut up. Suddenly— according to Duke: "The weapon went off." He'd meant only to *punch* the man, he'd claim, but had forgotten he held a loaded handgun. "It was such a mass confusion. I remember hearing the shot, but I don't remember pulling the trigger."

Shot in the head, the stranger cried out, and dropped. Eddie shrieked, too, in a voice of such pain and horror that Duke briefly thought he'd shot his friend.

Eddie may then have yelled: "You killed him! You killed him!" but Duke would say he heard, "Kill him! Kill him!" Duke stood above the wounded man and shot him point-blank in the head while Eddie took off running across a field toward distant woods.

"[He] was laying on the ground," Duke later said of his victim. "His eyes were still open. Mentally that's the biggest thing that I remember throughout the whole situation is his eyes."

A blue pickup truck approached, slowed down, and pulled over, and another local man, on his way home to pick up his son for baseball practice, walked toward what appeared to be a collision, with injuries. Duke felled him from a distance with a bullet to the head, stood over him, and shot him again, killing him, too.

In the woods, hearing more shots, Eddie had a moment of heart-stopping panic and wild indecision: Should he keep running, or return to Duke?

More scared of Duke than of anything else he could think of, he trudged back to the road and saw a second pickup truck, a second victim. Duke started shouting orders and, shaking violently, Eddie obeyed: collect the spent shell casings; wait for Duke to rifle through the men's pockets for their wallets; help Duke load the first man's body into the bed of his white pickup; drive it a few miles down the road, with Duke following in Eddie's car; pull over on the shoulder and abandon the truck; drive his own car with Duke in it back to Angel's; grab her wheelchair and their stuff and get her into the car; detour through the hamlet of St. Clairsville; pull over suddenly and run up into a wooded area to hide the wallets and spent shell casings; and speed back to Columbus.

The only other image Duke would recall clearly out of all the mayhem was Eddie squatting on the road, preparing to lift the first victim into the white truck, holding the stranger's legs, and crying.

They left behind a sloppy chronicle of bodies, bloody shoe prints, tire skid marks, and eyewitness views of the brown Charger's squealing, zigzagging flight around Belmont County. The police soon found the first victim "thrown face-down in the back of his pickup truck in the woods with two bullet holes in the left side of the head" and the second "lying beside his blue truck in the middle of the road [with] an entrance wound in each side of his head."

As Eddie drove west that night, Duke fell into a mumbling, drunken sleep in the passenger seat. Angel sat silently, wide-eyed, in the back. Eddie sobbed.

In Columbus, Duke told his ex-wife, Eddie's sister: "I'm going to prison for the rest of my life and I'm going to hell." Later Duke punched himself in the face so long and hard that the next morning he looked like he'd lost a fight.

Eight days after the homicides, Eddie and Duke were surprised in the parking lot of Duke's mother's apartment in Columbus and arrested. Duke was tried in state court in Martins Ferry. The murders were portrayed in court as execution-like rather than accidental. It seemed inconceivable that the violent deaths of two law-abiding local men represented nothing more than their bad luck running into two drunk punks from Columbus. The jury convicted Duke on all counts and the judge sentenced him to death.

Eddie Hill pleaded not guilty. At his trial, while there was no suggestion that he had killed anyone, evidence was plentiful that he'd been at the scene of the murders and had helped in every aspect of the cover-up. Police testified that it was Eddie Hill, not Donald Palmer, who had led them to the hidden wallets and spent shell casings in the St. Clairsville woods. The jury found him guilty of complicity in one count of aggravated murder, one count of murder, two counts of aggravated robbery, and two firearms specifications. Because he had not been the gunman, the judge gave him sixty years to life in prison, with parole possible after thirty years, rather than the death penalty.

IN MARCH 1990, EDDIE HILL, a humble, bewildered, mortified young man of twenty-four, moved into the Warren Correctional Institution (WCI). "I had no idea what to expect," he said. "Most guys coming in already had family or friends here or at least knew someone with prison experience to let them know what's up. I had none of that. I was over two hundred miles from home and all alone."

Hoping to get along with everybody, he was a trouble-free prisoner for the guards, but his eager-to-please manner with other inmates served him poorly, not unlike his past acquiescence to Duke. His parents made the six-hour round-trip drive on Sundays, so Eddie found himself rich in small change, snacks, and small material possessions that others lacked, but he shared them to a fault; he gave away everything he was asked for, and then was threatened with violence when he had nothing more to give. "My first big lesson here was: *Don't be nice,*" he told me. "The first thing I had to learn was how to say no."

Again, as in high school, he gravitated toward what he calls the "cool troublemaker types." He got yanked out of a fight by guards and sentenced to the "hole"—solitary confinement—for seventeen days. The loneliness was brutal. He emerged from it chastened rather than bitter. "I saw that I had to be distant, for my own safety," he says. "I became more wary."

Transferred to a new housing block after his release from solitary, Eddie steered clear of everyone. He had no idea whom to trust. All around him, men plowed through deep time alone, watching TV, exercising, and sleeping, watching TV, exercising, and sleeping, wrapping themselves up in the gray repetitive days. Eddie copied them. The monotonous routine ate up the flat days like a treadmill clocking miles. But then it struck him: the men on this TV/exercise/ sleep regimen weren't serving life sentences—their goal was to mow down shapeless time and cross it off. Unlike them, Eddie wasn't just scratching off weeks, months, and years; he would spend his *life* here. This was his *life*.

Hang out with a better class of people, he told himself. He began appraising the men around him and thought, *There are good people here and bad people here and many shades in between, just like on the outside.* One group gathered in the yard with guitars. He loved music, but had always assumed learning an instrument would be too hard for him. *I guess I have all the time in the world now,* he thought. He wrote to his parents to ask for an acoustic guitar, and they brought one the following month. In his cell, with a beginner's music book, he worked up some basic chords, then approached the men in the yard for informal lessons. They were open to him. No one called him "a natural"—that's not how men talked in here, but he saw their respect. His parents brought him an electric guitar when he asked. He joined the Music Association, and two rock bands.

He visited the prison library and checked out a book and read it from start to finish, and then another. He joined an in-house civic organization. He taught himself a new way to act around people—neither ingratiating nor distant, but something in between. His manner became calmer, less anxious, more dignified.

But it was a deeply lonely life, without sincere warmth or unguarded affection, without intimacy or physical contact. He'd grown up in a rural county, with dogs, horses, cows, and ducks all around. Other than his occasional glimpse of a red-tailed hawk banking over distant fields, he had entered a world without animals.

KAREN SHIRK COMPLIMENTED A FRIEND one day on a great shaggy mutt and asked where the dog had come from.

"Prison," said the friend.

"That's not a very nice way to refer to an animal shelter!" said Karen, but he corrected her: "No, I got him from Warren Correctional Institute. The prison. WCI. The prisoners train dogs for an animal shelter and adopt them out."

Prisoners training dogs? WCI in Lebanon, Ohio, was just thirty miles from Xenia. Could those prisoners train *her* dogs? The requests for dogs and the arrival of dogs outstripped Karen's capacity to house, care for, and train them all. She was so far beyond her original conception of *Maybe I could train four or five dogs a year for people turned down by the established agencies* that she didn't have a moment to think back about it, other than the obvious fact that far more than four or five families a year were being turned down by the major service dog agencies. Every 4 Paws dog's preparation required hundreds of hours of one-on-one work with a trainer. It would be a tremendous head start to begin *advanced* service training with dogs who'd mastered the basic obedience commands of "Sit," "Stay," "Down," "Come," and "Heel." As she began to imagine incarcerated people training rescued dogs to assist differently abled people, it struck her as an amazing win/win/win scenario, almost too good to be possible. But there turned out to be a world of animals behind bars in "prison-based animal programs," or PAPs.

THE SO-CALLED BIRDMAN OF ALCATRAZ was the first famous prison inmate to raise animals, though his name is a misnomer. In the 1920s, Robert Stroud, a violent psychopath and convicted murderer with a genius IQ, discovered a fallen nest of baby sparrows in the Leavenworth Penitentiary yard and raised them to adulthood. Over the next half-century at Leavenworth, mostly in solitary confinement, he delicately raised hundreds of sparrows and canaries, became a respected ornithologist, and published articles in scholarly journals. Meanwhile he stabbed inmates and guards and killed an orderly. Deemed too great a risk to everyone with whom he crossed paths, he was transferred to Alcatraz, where he spent the rest of his life without birds. In photos he is a creepy-looking guy who bears no resemblance to Burt Lancaster. He was the Birdman of Leavenworth.

In 1979 in Ohio, at the Lima State Hospital for the Criminally Insane, a male patient discovered an injured songbird on the hospital grounds, smuggled it inside, and cared for it surreptitiously. Over the next few days, other patients on the ward—silent, detached, uncooperative men—went out of their way to help him care for the bird. Advised of the situation, the director observed but refrained from intervening. Marveling at the sight of the glum, solitary, sometimes violent patients gently pulling together to care for the little bird, the director introduced fish, and then rabbits. Every new arrival kindled in the men the same gestures of benevolence and teamwork. When, after a year, the director compared behaviors of the men on that ward with the men on a parallel ward without animals, he discovered that the former required "half as much medication, had drastically reduced incidents of violence and had no suicide attempts during the year-long comparison." This was the first modern prison-based animal program.

A BIG FAT LONG-HAIRED FEMALE CAT somehow wandered one day in 1994 into the yard of Indiana State Prison in Michigan City, a maximum-security penitentiary for violent offenders. She was snatched up by an inmate and smuggled inside. Soon afterward, in an unrelated incident, a guard was murdered and the prison went into lockdown. Nine months later, when prisoners were permitted to communicate with one another again, the news spread secretly that the hidden cat had given birth and her owner was giving away kitties. Vigilant against a recurrence of violence, the guards sensed a conspiracy afoot. In searches of cells and bedding, they expected to turn up weapons, and found, instead, cats. And handmade cat toys. And twists of cloth used for feeding kittens drops of milk.

The warden ordered all cat owners to come forward and register their animals. If they did so, he promised, they could keep their pets, but any unregistered cats discovered after the deadline would be confiscated. Reluctantly, the men made themselves known. "To their delight . . . ," reported Stacy E. Smith in *PawPrints,* "the cats were allowed to stay and the pilot program at the Indiana State Prison began. Within a few years, there were 30 cats, and the doors were soon open to more." Even when the men were locked into the cells, the cats were free to slink in and out of the bars and visit other men in other cells and hobnob with other cats in the corridors. Their owners created elaborate climbing structures in their cells and, during outdoor periods, they walked their cats on leashes in the yard.

"These guys are really protective of the cats," corrections officers told a visiting cat behaviorist, Diana Partington. "There was a guy killed in here because he had spit soda pop onto someone else's cat." In general, however, the cats have brought greater stability and peace to the prison. Cat lovers in the vicinity of the prison reached out and now provide the prisoners with supplies, advice, and kittens, and the prisoners do the same for their neighbors.

Injured rabbits, songbirds, and raccoons are rehabilitated and released in Kentucky and Ohio prisons; pheasants are bred and released in Michigan. Horses, cows, hogs, goats, sheep, and chickens thrive in others. At the Mission Creek Corrections Center for Women in Belfair, Washington, inmates breed an endangered butterfly species, the flecked black-and-orange Taylor's checkerspot, recently releasing 3,600 caterpillars into the wild. At the Cedar Creek Corrections Center in Littlerock, Washington, inmates raise endangered Oregon spotted frogs for release into protected wetlands. That program has become the gold standard for herpetologists. "The frogs are entrusted to prisoners as eggs, not much bigger than the tip of a pencil," reported *Conservation* magazine. "Compared with frogs raised in zoo programs, the frogs at Cedar Creek are significantly beefier and reach maturity faster. Maybe that's because the prisoners . . . baby their tiny charges day and night. They monitor the water constantly and make sure that each frog gets plenty of crickets to eat. The prisoners keep detailed notes and growth charts as the eggs grow into tadpoles, then frogs, before being released into a protected wetland . . ." The success of the "captive breeding" program has caught the attention of zookeepers who now send their undersized frogs to Cedar Creek for special attention.

As a way for convicts to serve their communities, prison-based animal programs caught on across the country, not unlike prison-based license plate factories. But the animals turned out to do a great deal more for everyone than license plates ever did. PAPs appear today in nearly every state, in over two hundred prisons, and two-thirds involve dogs. One and a half percent involve cats and another one and a half percent, llamas.

WCI'S CAMPUS OF REDBRICK BUILDINGS, mown lawns, and curving walkways looks like a community college. In an all-purpose room, a 4 Paws dog-training class gathers; since 2000, 4 Paws has

been the only PAP operating at the Warren Correctional Institute. The arriving men—all dressed identically in long-sleeve
denim shirts and dark blue cotton trousers—pull plastic chairs
from a stack and arrange them in a circle. The youngest, a tall,
sweet-faced young black man with intricately braided cornrows,
wearing soft wheat Timberland boots, looks like he belongs at
Ohio State, University of Dayton, or Wright State—anywhere but
here. The oldest is in his late sixties, also African-American, a
towering man with weather-beaten eyes and isolated salt-and-
pepper tufts of hair and beard; his shirtsleeves swell across iron
biceps; his neck and wrists are encircled by tattoos. None of
these men mistakes this place for a community.

Around the perimeter of these forty-five acres on State Route
63 outside Lebanon, Ohio, stands a sky-scraping steel fence
topped by razor wire and surveilled twenty-four hours a day by
armed guards. Warren Correctional Institution is a Level 3 facility, in which 1,407 men requiring medium-, close-, or maximum-
level security are incarcerated. Today's students all have dogs
with them. Golden retrievers, black Labs, German shepherd dogs,
mixed breeds, and a few dainty high-stepping papillons promenade into the room beside the somber men. The inmates stand
quietly and speak in low tones—each has slowed his inner clock to
match the imposed monotony here—but the happy, well-groomed
dogs, gleaming with life, have made no such adjustment.

The inmates can't help exchanging glances and frowns of
paternal pride. "How's your Meatball?" a fellow will say and
then, gazing down at his own dog, mention, "Yeah, this old gal
picked up the commands pretty quick."

In their former lives, the men in the 4 Paws class committed aggravated robbery, property offenses, or murder; but no
one here was convicted of domestic violence or child abuse—

those categories of offenders are excluded from virtually all PAPs everywhere. "Domestic abusers are known as a group to be dangerous to animals," Karen Shirk told me. The data is overwhelming. According to the American Humane Society and the National Coalition Against Domestic Violence, 71 percent of women who took refuge in shelters reported that their batterer had threatened, injured, maimed, or killed a family pet for revenge or for psychological control. According to the ASPCA, "Between 18 and 48 percent of battered women, and their children, delay leaving abusive situations in fear for what might happen to their animals."

So WCI's 4 Paws class is not open to domestic abusers. But those convicted of other categories of crime, including murder, may apply to the program and, once accepted, may stay in it as long as they demonstrate respect and kindness toward the dogs in their charge. The men make a bit of money, acquire real-world job skills, and exit prison with letters of recommendation and career thoughts. The benefits run deeper: participants in prison animal programs express greater feelings of self-worth, emotional connection, and sense of purpose in life. Those positive traits lead to reduced rates of rule-breaking and violence inside the institutions, and reduced rates of criminal behavior and recidivism after release.

AND HOW DOES THAT WORK exactly?

For those incarcerated individuals lucky enough to work with dogs, it works the same way as it works for the rest of us. There's something about spending time with a dog that makes everything in life a little sweeter. Though we've grown accustomed to it—though we've come to expect it—it's extraordinary how dogs put their lives in our hands. Not just their muzzles, which they willingly lay across our palms if we ask—twitchy, quilled peninsulas of fur, outlined with sly toothy alligator smiles—but their whole lives. Most have no choice but to accept the new human keepers who lead them, leashed, away from their place of birth or rescue, and require them to adapt overnight to completely new lives.

Dogs could be sulky about it; they could drag their feet when asked to cross a new threshold. They could refuse to deferentially greet the elderly dachshund napping on the rug with his short legs out in front of him or the long-haired cat peering down coldly from atop the piano. They could cower under the kitchen table at the approach of our big shoes. They could give a sad lick of the lips and reject the food we offer. They could decline to sleep on our beds. They could maintain a vigil—standing tippy-toe on a chair by a front window—in tireless watch for the person who handed them over to this strange fate where nothing smells familiar. But that's not how dogs act. Most, within hours of meeting us, are making eye contact, offering hopeful wags of the tail, even hopping up—if invited (or not)—onto our beds. Though they've hardly gotten their bearings, they have already implicitly promised, being dogs, to do their best.

Scientists are interested in the underpinnings of why humans feel happier in the presence of dogs. In 1995, researchers at the University of Maryland Hospital found that "heart attack patients with dogs were eight times more likely to be alive a year later than people without dogs." In 1999, SUNY's University at Buffalo followed twenty-four stockbrokers taking medication for high blood pressure and found that "adding a dog or cat to the stockbrokers' lives helped stabilize and reduce their stress levels." In 1999, Swedish researchers found that "children exposed to pets during the first year of life had fewer allergies and less asthma."

In 2004, researchers at the University of Missouri in Columbia, building on earlier discoveries made by South African scientists, measured the blood pressure and hormonal levels of volunteers across three scenarios: they petted and played with their own dogs, they petted and played with friendly dogs they didn't know, or they petted robot dogs. The people interacting with their own dogs experienced a drop in blood pressure and a cascade of hormonal changes, including the release of serotonin, prolactin, and oxytocin (the "feel-good" hormones) and a decrease in cortisol, the primary stress hormone.

Interacting with unfamiliar dogs relaxed people, too, though not to the extent they'd experienced in the company of their own dogs.

Petting and talking to a robotic dog, rather understandably, caused everybody's stress levels to rise. I feel unhappy just thinking about it.

"Elderly people, living alone, can be saved from depression caused by loneliness, or feelings of uselessness, when they share their lives with a beloved cat or dog," Jane Goodall has written. "This is not just because animals are soft, furry, and warm. It is because these animal healers seem to empathize with their humans, understand their needs—and love them . . . A mechanical stuffed toy animal, no matter how skillfully crafted, no matter how lifelike it appears, will never take the place of a living, feeling, and loving animal."

Scientists in Japan just announced the discovery that dog owners experience a surge of oxytocin simply through prolonged *eye* contact with their dogs. Oxytocin is the hormone of maternal love: it spikes in mothers who gaze into the eyes of their children, enhancing their caregiving urges and tightening the sense of intimacy. "In one sense, love is a name for a feeling that evolution uses to trick us into performing risky, costly behaviors such as child rearing and the defense of our mates and children," writes Carl Safina. "Love helps commit us to them . . . That doesn't mean love is not profound. It only means that love grows from a deep tangle of roots." Dr. Takefumi Kikusui, of Azabu University in Sagamihara, Kanagawa, Japan, lead author of the study, suggested that finding this mechanism at work not only in humans gazing at their babies, but in humans gazing into the eyes of their *dogs,* points to a truly ancient biological connection between our species.

Dr. Alan Beck had foreseen as much: "The warm feeling we get from our dogs and other pets isn't just a learned behavior, but something that's hard-wired into humans . . . an inborn attraction to nature."

Clearly this inborn attraction to nature—of which dogs are the greatest ambassadors in the modern era—blooms within humans of all ages and in all stages of sickness or health, cognitive development or decline. It's true for people in daily contact with all sorts of creatures and it's truer still for those unfortunates walled away from the animal world. "When people face real adversity . . . affection from a pet takes on new meaning," write Alan Beck and Aaron Katcher in *Between Pets and People: The Importance of Animal Companionship.* "The pet's continuing affection is a sign that the essence of the person has not been damaged."

"I'm not as stupid as I was always told I was," said an inmate enrolled in a PAP with dogs. "I've been complimented by officers who had been tough on me before," said another. "Being an inmate doesn't make us evil," realized a third.

"Human subpopulations that have been previously ostracized or considered deviant by the dominant culture, including people with disabilities and those institutionalized in prisons and hospitals . . . may be particularly able to benefit from the unique, nonverbal type of interactions that take place with animals," writes criminologist Gennifer Furst. "Without language to offend or cause harm, interactions between people and animals can feel less judgmental and therefore more therapeutic for incarcerated people."

Karen Shirk says: "I've seen convicted felons—including murderers—just melt and cry when it's time to give back their dogs."

4 PAWS PUPPIES ONCE LIVED at WCI for three months, but Karen, Jeremy, and the trainers sadly realized that the dogs came back from prison "institutionalized." They shied away from crowds and traffic and knew nothing of children; they could barely cross a street or a park. It took months of gentle socializing by trained foster families to help the dogs overcome their fears, and some never recovered sufficiently to prepare for service. They had to set a maximum of two months behind bars for their puppies.

Since most 4 Paws dogs are trained for children with disabilities—and children are more interested in having playmates and friends than in having four-legged nurses, therapists, or trackers—Jeremy asks the prisoners to teach the dogs tricks: "Roll over," "Speak," "Gimme five," and "Play dead." "I had learned with Ben that a dog helps you make friends," Karen told me. "We place dogs with kids in wheelchairs, kids on ventilators, kids with autism, kids with spina bifida, kids with dwarfism, kids with seizure disorder, kids with cognitive impairments and hearing impairments—it doesn't matter—if your kid has a dog that does tricks, other kids will come over to make friends. Kids will ignore a disability if you've got a cool dog."

One prisoner, with a sense of humor, returned a dog with a novel response to the command "Play dead." The dog *lurched*, as if shot, staggered across the floor, knelt, got up, knelt again, whined piteously, and then dramatically collapsed and closed his eyes. Cool dog. Lucky kid.

As the men recall their dogs to heel, the tall, powerful-looking tattooed man with the weathered eyes bends over and swoops up a petite, silky little thing, a fawn-and-white papillon, eight pounds of fluff. He strokes her long fur and suddenly announces loudly to the class in an improbable falsetto voice: "Say: 'Hello everyone!' Say: 'I've been practicing so hard for my class today!'"

CHAPTER 13

Lucy & Jolly

Military veterans returning from Iraq and Afghanistan are presenting symptoms of post-traumatic stress disorder at epidemic levels. In the search for relief, many are exploring alternative treatments, including yoga, acupuncture, herbal remedies, massage therapy, and service dogs. PTSD service dogs are trained to sense when their handlers are in the grip of a flashback, panic attack, or nightmare, and to prod them out of it, recalling their humans to present-day Birmingham or Tulsa and away from a horrifying memory of Fallujah or Samarra; they may be asked to perform tasks like searching a dentist's office for mines, to assure their handler that it is safe to enter. "Veterans rely on their dogs to gauge the safety of their surroundings, allowing them to venture into public places without constantly scanning for snipers, hidden bombs, and other dangers lurking in the minds of those with [PTSD]," reports Janie Lorber for *The New York Times*.

Like all dogs, psychiatric service dogs need daily walks and outings, so the handlers—no matter how depressed or anxious—must get out of bed, get dressed, and go outside. Dogs enjoy meeting new people and other dogs, which tends to entangle their owners in eye contact and conversation. PTSD dogs, just by being dogs, drag their humans into everyday life, while offering constant reassurance that it's safe there. "Skeptics say that dogs cannot possibly treat the underlying disorder, where memories of traumatic events trigger potentially debilitating symptoms," writes James Dao in the *Times,* "but many PTSD experts say that there is much anecdotal evidence that dogs make veterans feel better—and that may be enough. 'If the point is to treat a person into remission, we have no evidence that service dogs can do that,' said Alan L. Peterson, a professor of psychiatry at the University of Texas Health Science Center in San Antonio . . . 'But in terms of just coping, they might help.' "

When Eleanor Keith heard about PTSD service dogs, she researched their appropriateness for children and discovered 4 Paws for Ability. In August 2011, she emailed Karen Shirk with the subject line: "Wondering if we even qualify," and began: "My daughter is a traumatized, attachment-disordered child." Karen Shirk emailed back, "ABSOLUTELY." 4 Paws had been placing dogs with individuals with attachment issues and with PTSD since 2005, six years ahead of the U.S. military.

In December 2012, after just four months of fund-raising, the Keiths drove to Xenia.

SEATED IN THE TRAINING CIRCLE with other families, James and Eleanor, in a state of almost painful suspense, flashed back to the waiting room of the Chinese orphanage in which they had prepared to meet their daughter. They sat up straighter when their dog—a yellow Lab named Jolly—hurtled into the room. "Lucy, look! It's your dog! It's Jolly," they said, exhaling in relief, hugging the child, but she weaseled out of their arms and went to stand behind the father. She hadn't asked for a dog. The situation raised a few urgent questions: *What if the parents like the dog more than they like me? Why do they need him? Why am I not enough for them? What if the dog doesn't like me?*

The other parents laughed warmly when Jolly galumphed across the circle beside his handler to meet his new family. He was young and gangly, sloppy and happy. He didn't like the Keiths specifically—he didn't know them—he just liked everybody! Soon, if all went well with his training at home, other people would fall away and he would narrow his love to the three Keiths, especially—if all went *very* well—to Lucy. He slobbered over James's and Eleanor's hands, eagerly licking up the treats they offered. While the treat bags were distributed, Jeremy had said: "Your *child* should be the one giving treats. If your child is unable to reward the dog, then it's your job to hide the treats on and around the child. You want the dog to feel:

I don't know what it is about this kid, but whenever this kid is around, good things happen."

Guided by Eleanor, Lucy's small hand now appeared from behind James with a treat held upon her open palm. But she didn't like the dog's drippy tongue and withdrew quickly.

James liked Jolly instantly. He had nothing against papillons, but he could hardly believe his luck in getting this good, basic dog. It was almost like doing a regular thing for once. Jolly looked like a dog James would have brought home from a shelter, a dog who would ride shotgun in the pickup truck and grin as the wind flapped his lips and ears. But . . . was this the right dog for his anxious girl? A quick side-glance told James that Eleanor was already concerned.

"We give you young dogs," Jeremy Dulebohn was explaining. "Service dogs for *adults* can be placed at older ages. We give you young ones so the dogs can grow up with your children. Kids are full of noises and behaviors that can scare dogs who aren't used to them. Once a dog gets spooked by a child, it's hard to reverse that impression. So we place our dogs as young as possible and they get habituated to your children. But it means you may see some puppy behaviors. You may see a lot of chewing and teething."

Jolly, relaxing at their feet, began to snake toward the nearest dog—one of his best buddies!—and the two, staying low because they were supposed to be "Down," began mouthing each other's muzzles. *Mmm—mmm-mmm.* Then they forgot "Down" and the class and everything except the desire to roll around on the linoleum floor like wrestlers. Jeremy stopped class to help the two families gently recall their dogs. *He's a big puppy!* Eleanor realized. Followed by: *They're sending us home with a puppy? I'm supposed to finish training him?* Again, it seemed, the Keiths had gone to great lengths, traveled far distances, and spent unthinkable sums of money in pursuit of . . . the average—bringing home a yellow dog—with the risk that they might fail at it.

"He's too playful and too strong for Lucy," Eleanor said during a coffee break.

James nodded and dutifully crossed the room to have a word with Jeremy, offering to swap out the dog he already liked. Duty had long since displaced expectations of happiness in James's life. When Jeremy assured James that Jolly was the right pick for their family, James returned to Eleanor's side with a shrug of resignation, but he wasn't entirely disappointed.

THERE ARE 4 PAWS FAMILIES who bond with their new dogs quickly. There are children who throw themselves across their furry friends the moment they're introduced. This experience would not be the Keiths'. Eleanor was trying to follow Jeremy's instructions about letting the child give treats to the dog but, unlike some of the more challenged children in the training circle, Lucy saw exactly what Eleanor was trying to do and did *not* want dog treats hidden in her pockets or around her feet. She wasn't going to bribe the stupid dog to like her! She pushed the treats away with an attitude of *He's probably not going to like me anyway* and escaped back to the play area.

Eleanor began to panic. That night at the hotel, there was a knot in her stomach that felt like buyer's remorse or even like the melancholy state of mind known as post-adoption depression. "I was only six weeks out from my sister's suicide," Eleanor told me. "Focused and calm I was not."

James was fending off something in his stomach, too, and it turned out to be the flu. He got into bed, feverish. "It'll take time to get to know Jolly, don't you think?" Eleanor called to him above the noise of Lucy's tantrum. Lucy flipped out for a long time that night, her wails alternately clear and muffled as she rolled around, kicking, on the motel bedspread.

After ten days in Xenia, the Keiths drove 612 miles back to Iowa in their Chevrolet Impala. They rode mostly in silence, conflicted. James had stayed in bed sick for five days, missing half the classes. Eleanor had sat through every minute, her perceptions clouded by worry and exhaustion. Now they both feared they were transporting a hugely expensive uncontrollable sixty-five-pound mistake across state lines. In the backseat, Jolly looked out the window or gnawed on his Nylabone or licked his front paws or snoozed. Whenever James or Eleanor turned around to greet him, he sat up whacking his tail against the seat. When they called, "How's your new dog doing, Lucy?" she made no reply. Lucy pressed herself against her door as far away from the big dog as possible.

NO ONE WOULD EVER AGAIN gallop through their front door prepared to pay such close, devoted attention to Lucy as the shaggy fellow from 4 Paws. In the front hall he shimmied and threw off sparkles of snow from the front yard, then bounded up the short staircase to the living room. He sniffed in the direction of the old cat, who stiffly exited. He bent to investigate the elderly shih tzu, Murphy, whose flattened face of down-flowing fur wore a permanent look of *Oh, good grief,* now *what?* Released into the backyard, Jolly tracked along the aluminum fence, looked up at birds, squatted, took a steaming dump in the snow, and ate it.

Back inside, he slid down like the Sphinx on the kitchen floor— front legs extended, head nobly lifted—and waited for the next interesting thing to happen. Jolly was a happy, optimistic guy.

But complicated emotional terrain awaited him. Here were three human hearts confusingly stalled—starting toward him, stopping, and lurching away. Lucy made a great show of disinterest. She ran to her room and slammed the door. She didn't have a handle on the parents' intentions: *What if they love the dog more than me? What if the dog hates me?*

Eleanor followed, worrying that they were blowing the homecoming. "Honey? Jolly's looking for you," she said in too pleading a voice from the hall. Lucy didn't reply. It was obviously a lie. Lucy was right: Jolly was not looking for her.

In the living room, James sat down on the sofa. "Come here, boy," he said. Jolly came and stood facing the man, and then inched forward. James kneaded the dog's supple, kid-glove ears. The dog moaned in pleasure.

But then the man remembered that Lucy was supposed to be the dog's first love, so he got up to move the suitcases down the hall. Jolly tagged along. He was surprised when James suddenly stepped into a bedroom and gently closed the door against the dog. Now Eleanor and Jolly both stood in the short hallway facing a closed door, like two unacquainted restaurant guests politely waiting their turns outside the restrooms.

That night, Eleanor invited Jolly to hop up onto Lucy's bed. Dogs from 4 Paws were legendary among parents for helping children (and parents) sleep through the night. The longing, after seven years, for an easier bedtime process and for a night of uninterrupted sleep scraped against Eleanor's eyelids from inside, so she pressed on despite Lucy's objections. *I don't know if he wants to sleep with me or they're just making him sleep with me,* Lucy may have been thinking. *They might want him to sleep with me so they won't have to lie down with me anymore at bedtime!*

But the well-meaning dog was foiled by the twin bed with the slippery satin bedspread. He couldn't seem to get his footing. He jumped up and slipped off, jumped up and slipped off. He gave up with a sigh and settled on the floor beside the bed. Lucy huddled against the headboard with her knees drawn up and watched with an odd little half-smile as the dog jumped and slipped and failed. It might not have been that she enjoyed watching him fail; perhaps his failure (*I fail at everything!*) might have made him a little more of a person to her.

In the small hours of the morning, Lucy woke up with night terrors as usual. When Eleanor staggered into the room, Jolly sat up on the floor and looked concerned. "Oh, Jolly," Eleanor murmured, disappointed, and he woefully thumped his tail on the floor.

Lucy left for school on Jolly's first morning without telling the dog goodbye. "She's not a special education student who is provided with an aide or dog-handler at school," Eleanor told me. "She works at or above grade level. She has no behavioral issues there—she'd be too ashamed to act out at school." So Lucy left and Jolly spent the day with Eleanor.

When Lucy came in the front door at three o'clock, Jolly was thrilled to see her. Already accustomed to being showered with treats when Lucy was around, Jolly responded to her with Jeremy's prescripted feeling, *I don't know what it is about this kid, but whenever this kid is around, good things happen!* But Lucy yelled, "I don't want him!" and ran to her room.

"As usual, she felt trapped," Eleanor told me. "She felt pressured by us to succeed in bonding with Jolly, so everything about it became, in her eyes, another set-up for failure. If we'd never followed through with our idea and gone to 4 Paws in the first place, she might have thought: *I don't deserve a dog.* If we'd given Jolly away after her rejections, she might have thought: *He didn't like me enough to stay.* I think she wanted to make friends with him, but anxiety ruins everything."

JAMES AND ELEANOR HAD FANTASIZED that Lucy would let the dog into her affections. After all, no *dogs* had disappointed or abandoned her in China. On the other hand, Murphy the shih tzu avoided her. "Lucy even worries about whether animals judge her and reject her," Eleanor says. "And, actually, it is true that Murphy doesn't like her. He's deathly afraid of her meltdowns and will threaten to nip if she gets too close."

Now the parents watched helplessly as Lucy drew boundaries around herself so gerrymandered that neither Jolly's friendly approach nor Jolly's indifference suited her. Evidently, for a child with attachment issues, *no* one's love is welcome, not even the love offered by a handsome young pure-hearted pup without a nip in him. In this house, you had to prove you loved Lucy a thousand times before she would trust your love. And even then she wouldn't trust it.

JOLLY WAS FULL OF SELF-CENTERED DELIGHTS. People like to say that what they love best about dogs is their loyalty, but the opposite is also true. We love dogs' self-centeredness, and their impression that our lives revolve around them. And it's undeniable that, no matter what people struggle with day-to-day, there is great solace and pleasure in being able to arrange a dog's life for maximum happiness.

A well-cared-for dog, a contented dog, convinced that he or she is the center of the universe, is a great sign that you're doing something right. Jolly represented a new chance for the Keiths to do one simple thing right.

Like most dogs, Jolly could probably tell who among his new human and animal acquaintances felt love for him (James); who felt confused (Eleanor—"I felt responsible for this project; I desperately wanted this experiment to succeed," she told me); and who felt fear, active dislike, or ambivalence (old cat, elderly dog, Lucy). He turned, wagging and smiling, toward James. James's return from work in the evening was the occasion of Jolly's greatest glee. The dog did a quick tap-dance routine, banged his flanks against James's legs, and opened his jaws wide to mouth James's hands. Too soon, the man reluctantly, self-consciously separated himself from the dog, ending the display.

There was no reception like that for Lucy after school. When Jolly approached, she was unwelcoming, rather neutral, on edge. *I don't know what it is about this kid, but* . . . would have been extinguished by now, had not Eleanor quickly supplied treats to Jolly in order to preserve the rest of the thought . . . *but whenever this kid's around, good things happen!*

PSYCHIATRIC SERVICE DOGS WORK on a landscape as treacherous in its way as the urban traffic navigated by guide dogs. Instead of speeding cars and roaring buses, dangers appear in the form of human twitches, racing pulses, rising blood pressure, sudden stillness, shallow breathing, or tantrums.

Jolly watched Lucy and took in all the signs as she careened like an out-of-control bus down a mountain pass through high anxiety, panic, and rage, but he seemed unaware that it had anything to do with him. He had been trained in "Nudge," "Over," "Lap," "Kiss," and "Touch." The theory was that Eleanor or James would give one of these commands when Lucy hit a rough patch and that she, having fallen in love with the big lug of a dog, would be calmed by his touch.

But Lucy had not fallen in love with Jolly and didn't want him touching her. She seemed deliberately to interpret "Nudge" as aggressive. She yowled and lost her balance and kicked to keep him and everybody at bay. She was alone, alone, irremediably alone!

Eleanor told me, "When she started melting down, I'd tell Jolly, 'Get your girl!' or 'Kiss!' in a happy, excited voice, which ran counter to the feelings being expressed by Lucy. I think Lucy interpreted my enthusiastic encouragement toward Jolly as a rejection of *her*. She didn't want me to cue him or reward him—it all just made her madder."

With misgiving, Eleanor continued to reward Jolly whenever Lucy threw a fit, so that, unlike Murphy the shih tzu, he wouldn't flee. But this left Jolly's conditioning slightly off—he was getting rewarded for being a bystander to Lucy's meltdowns. *I don't know what it is about this kid, but when she goes ballistic, good things happen!* Eleanor couldn't seem to connect the dots for Jolly that his job was to stabilize the child. She told James: "I say 'Nudge' so he'll interrupt the meltdown, but his 'Nudge' is too strong for her—he knocks her over."

"So don't say 'Nudge,' " he said.

"Right," she said uncertainly.

When Lucy came unglued the next afternoon, Eleanor didn't say "Nudge," Jolly didn't nudge, Lucy continued to rage, and Jolly beamed up at Eleanor with an expectant smile and a wag, as if everything was unfolding according to plan. He continued to seek eye contact with Eleanor until she remembered to give him his dog treat, which he swiped with a slobbery whisk of the tongue and took under the dining room table to enjoy.

Jolly had "Nudged" Eleanor. Eleanor was getting so well trained!

"A PAW ON THE LEG. A nose nuzzling against your arm. Maybe even a hop onto your lap. Dogs always seem to know when you're upset and need extra love, even though they hardly understand a word of what you say," reported Michaeleen Doucleff for NPR in 2014. "How can that be?"

Dogs have a voice-sensitive region in the brain similarly placed to that of the voice-sensitive region in the human brain, research has found; moreover, dogs and humans respond to vocalized emotional clues similarly. "The happier the barks or giggles, the more that brain region lights up," reported Doucleff. "The sadder the growls or whines, the less it responds."

"Like people, dogs use simple acoustic parameters to extract out the feelings from a sound," said Dr. Attila Andics, a neurobiologist at the MTA-ELTE Comparative Ethology Research Group in Budapest and chief author of a study that involved scanning the brains of dogs and humans as they listened to similar dog and human sounds. "For instance, when you laugh, 'Ha ha ha,' it has short, quick pieces. But if you make the pieces longer, 'Haaaa, haaaa, haaaa,' it starts to sound like crying or whining. This is what people—and dogs—pay attention to . . . This is a first step to understanding how dogs can be so attuned to their owner's feelings."

When Eleanor handed Jolly a biscuit for being in attendance while Lucy raged, Jolly accepted his prize with lowered ears and an appreciative lick, possibly sensing that Eleanor had handed him the treat dutifully rather than enthusiastically. When she told him, with a sigh, "Good boy," Jolly might have intuited the chagrin behind the mixed message of: *You're a good boy, Jolly, I guess this is the best you can do.*

When James stood still in the foyer after work for the dog's homecoming displays, less reciprocal than he had been in the past, Jolly sometimes slowed down briefly. Or sometimes he rejoiced even harder! *This is how happy I am to see you! This is how happy!* But the man's heart grew small. He told Jolly hello, but never sat down and toyed with the dog's ears anymore.

EAGER TO PLEASE JAMES AND ELEANOR, Jolly began adjusting his responses. Every once in a while, Eleanor fed Jolly a biscuit as a heartfelt gift emerging out of happy feelings, with quick words like *Good work, Jolly! Thataway, boy!* Those were the biscuits he wanted! He wanted to hear the short happy sounds.

One afternoon, after finishing a turkey-flavored biscuit he earned for being present at Lucy's afternoon tantrum—"Here you go, Jolly," Eleanor had said in a lower and slower voice—Jolly noticed that Lucy was still screaming and that Eleanor was still distressed. The screaming didn't bother him—*I don't know what it is about this kid, but when she goes ballistic . . . !*—but Eleanor's sorrow probably unnerved him. He swept the last crumbs from his lips and approached Eleanor to place his muzzle in the curve of her hand, but she withdrew her hand and used it to brush back a strand of hair. She was sad! Or mad! Was there something he was supposed to do here? "Nudge" had been extinguished as a command, but something of the notion of physically helping Lucy may have lingered.

He flattened his ears, lowered his head, gave the ground a couple of sniffs, and gently approached wildly out-of-control Lucy, relaying through dog body language that he meant no harm, that he was a friend. He stopped and pushed his head toward her hands, which she alternately wrung and waved while shrieking. When one of her tense small hands flew by his face, he gave it a lick. The sensation distracted her for a second—there was a break in her bellowing, as if a record had skipped. Eleanor, with her back to them, heard the skip and turned around. Jolly caught the small flapping hand again with a second lick and Lucy gave a snort of surprise. When he tried to catch her hand again, she giggled. Even though she pulled her hands away and waved her palms at him to stop, Jolly took the giggle as a green light. He extended his neck and tagged with his nose one of the palms vibrating in protest in front of him. It was almost like . . . play. The girl and the dog looked straight into each other's eyes and Lucy stopped screaming. Eleanor caught her breath and then gave a soft chuckle. Her moment of pleasure surely came through loud and clear to Jolly, a far greater reward than a biscuit.

That night she told James about the moment that Jolly had interrupted Lucy's fit, and he made a short noise of surprise and looked over at the dog. Jolly sat up straighter and his eyes brightened.

Soon there came the famous type of 4 Paws morning Eleanor and James had only heard about: they woke up in their own bed, having slept through the night. This meant that when Lucy cried out with night terrors, Jolly must have cuddled her back to sleep. It was the Keiths' first uninterrupted night of sleep in many years, perhaps since they'd become parents.

ELEANOR WAS THE FIRST TO NOTICE that Lucy was calling Jolly "my dog." She played hide-and-seek with him. Eleanor told Jolly, "Stay," and Lucy raced off down the hall, turned into the bathroom, and lay flat in the bathtub, mirthful at the thought that he would never find her in the tub. When eight-year-old Lucy called his name, Eleanor released him, and he hurtled straight to Lucy. "Mommy, he found me!" squawked a little voice quickly muffled by dog kisses.

A blizzard fell so thickly one day that it closed even *Iowa* schools. Lucy ventured into the icy wilderness of the backyard with Jolly. He plowed tracks with his face and she tromped behind him. When she looked up through her steamed-up glasses and saw her parents watching, flabbergasted, from the kitchen window, she laughed. Jolly dug down, found some old poop he'd previously overlooked, and crunchily ate the frozen lumps, making Lucy laugh harder. Back inside, Eleanor knelt to pull off Lucy's snow-packed boots while James heated up soup for lunch, the adults reveling in every quiet moment of this astoundingly normal Snow Day. Lucy liked Jolly's rough coat now. She groomed him with her fingers; she kissed him. She played restaurant with him, propping a pretend menu in front of him, asking him for his order, and then bringing odds and ends of food from the kitchen while he waited politely. One day that winter, Lucy looked up, thought a moment, and announced: "I just went seven days without a meltdown."

In July 2013, the family drove to Niagara Falls. It was a two-day road trip and Lucy tolerated it. Overlooking the falls, the Keiths dressed identically in blue plastic rain ponchos. In every photograph of the trip, Lucy is in physical contact with her parents or her dog; she is cuddled, held, touched. Her dad holds her hand, her mom drapes an arm around her shoulders, and Lucy holds on to Jolly's fur so he doesn't fall over the waterfall. When it was time to go home, Lucy double-checked: "Jolly's coming, too?"

"Of course Jolly is coming! Jolly is part of our family," said Eleanor.

Lucy might then have reasoned along the lines of: *Jolly is my dog. If Jolly is part of the family, then I am part of the family. If they're always going to keep Jolly, then they're always going to keep me.*

Lucy's revelation freed James to befriend Jolly again. Lucy had moved beyond her suspicions: *They love him more than they love me! He loves them more than he loves me!* "Come here, boy," said James in the living room. The dog lowered his head respectfully and approached. James picked up the warm, complicated ears and rubbed them and the dog shivered with pleasure.

In February, the three Keiths pulled on matching purple University of Northern Iowa sweatshirts and went to cheer the Lady Panthers basketball team. Jolly came, too. Lucy had learned to channel her anxiety into concern for Jolly—"Don't worry, boy.

You'll be all right, boy," she told him in the backseat in a high, trembly voice. It was Lucy's first time at a sporting event—the sensory overload of such an event would have been unthinkable even two months earlier. Sitting between her parents, cheering when they cheered and yelling "Ohhhh!" when they were disappointed, dressed like a miniature version of them, there was no mistaking her for anything other than their daughter. The look on James's face—a father seated at a basketball game beside his daughter and wife, with a good yellow dog at his feet—was neither pained nor brave; he looked more than happy; he looked wistfully happy: immersed in the event, while keenly aware that it was the kind of everyday happy family moment he'd thought would elude them forever.

CHAPTER 14

Casey & Connor II

It was their Thousand Days. It was the Millard family's Camelot. Deb and Scott and Connor and Casey: a family of four. They went on walks, and to restaurants, and to the mall, and (once) (never again) (too much sand in the medical apparatus) to the beach. Standing in the kitchen, Connor helped

Mama make pizza or cookies while Casey watched the floor for crumbs. With his grandpa, Connor rooted for the Ohio State Buckeyes. When Dick Tiel cheered, Connor threw up his arms to signal, *Touchdown!* His bedroom was no longer a nursery; it was a big boy's room decked out in Buckeye black-and-red, with OSU curtains and sheets and posters and a rug shaped like a football.

Their "Thousand Days" as a foursome actually only lasted about 750 days. The first 180 were filled with Casey's romping and Connor's breathy laughter; the rest put them on a steep decline into darkness.

In the late fall and early winter of 2005, the Millards' intuition—that the sinking of Connor's spirit corresponded to unseen physical deterioration—recurred. Unaware that the rules were changing, Casey paddled hard to keep her boy afloat and happy. Wearing her red 4 Paws service dog vest, she hopped up into the backseat of the minivan at all hours of the day or night for trips to the emergency room. Connor, feverish, in distress, flopped a limp hand over the edge of the car seat. Casey did as much as she could with the hand, the way she frolicked around the barely tossed tennis ball in the backyard. She licked the small hand assiduously, front and back and between the fingers; and she bumped her head against Connor's palm from below, helping Connor to pet her; and she pretended to gnaw on his wrist. If Connor gave an audible little snort, Casey took that as laughter and whacked her tail in reply. *We're having a fun time!*

As Deb or Scott carried Connor through the hospital doors, Casey cantered beside them and rested on the floor near the wall of whatever examination room they were assigned. She refused to leave Connor. She disobeyed if Deb or Scott, exiting, said, "Come," if it meant leaving Connor. She refused even to go outside for a walk to relieve herself; Scott or Deb had to drag her. The hospital staff respected Casey. In X-ray, the radiologist draped a lead blanket over her and let her remain in the room with the patient. But, as medical procedures grew rougher and more invasive, Casey evidently sensed Connor's rising fright and agitation. She made rumbling noises at doctors and nurses. Then she tried to stop them from coming into the room. At first the medical personnel laughed and said, "It's okay, Casey, we're not going to hurt your boy," and detoured around her. But she didn't take her signals from them; she took them from Connor and Connor didn't want these people here. Then Casey stood up and tried to block their approach to Connor's bed. She lowered her head and growled. Then she bared her teeth.

"Okay, that's it, that's enough," said Scott. He attached the lead. "Casey, come!"

"No! Ay-ee! Ay-ee!" screamed Connor.

"Casey, heel!" said Scott, and led her from the room.

After that, Casey was left at home when they made the hospital runs. It was not easy to get out the door without her. A few hours later, when Scott or Deb returned to feed her and let her outside, they found her pacing between Connor's room and the front door.

The news wasn't good in the winter of 2005 to 2006. With more sophisticated DNA tests now available, doctors had finally diagnosed Connor. He had mitochondrial disease, a rare genetic mutation that caused the energy-producing organelles within his cells to dysfunction. Connor's type would lead, in time, to global organ failure, one system at a time. Now he was plagued by stomach, intestine, and liver problems, including the inability to digest solid food; bacterial and yeast infections in his catheters; fevers and low blood counts. He spent a week in intensive care between Christmas 2005 and New Year's 2006. The Millards brought Casey in for a visit, after extensive warnings and reminders to be good. She bounced up onto his hospital bed and found her way—like she always had—under or through the wires and tubes, into his arms.

Seeing Connor come to life when Casey arrived, Scott thought, *Connor could outlive us.* But sometimes he wondered if the equipment gave them a false sense of security. He seemed stable, but was that a mechanical illusion?

Connor was sent home in January with the digestive issues unresolved; he could no longer eat solid food and, on formula alone, his weight and energy plummeted. He turned seven on March 1, 2006, weighing twenty-three pounds. He returned to the hospital for the insertion of a G-tube into his stomach for liquid food delivery, but suffered intense stomach pain. During the next hospital stay, a central IV line was installed in his chest for delivery of nutrition, but still he wasted away. The doctors detected significant immune deficiencies: Connor's bone marrow wasn't producing enough blood cells. They gave him blood transfusions and iron transfusions, but he continued to deteriorate.

There were tragic markers of regression for the Millards. Connor walking, Connor feeding himself, Connor shot-putting a tennis ball to Casey in the backyard now stood revealed as the lifetime summits of his physical achievements. His preschool concepts—a few letters, a few animal names—now appeared to represent the high points of his intellectual attainments. A child with degenerative disease loses abilities, loses thoughts. At seven, he functioned more like a four-year-old and was moving backward, capable of less every day.

On the red-letter days when Connor came home from the hospital, Casey stood on her hind legs looking out the front window—predictive, intuitive about the boy's impending arrival. She dropped from the window to run to the front door, then back to the window, then back to the door. She didn't bark (Connor didn't like barking—she never barked in his presence) but, as Deb unlocked the door and Scott carried Connor over the threshold, Casey was a one-girl brass band, wagging and wiggling and jogging in place, her static-filled kinky pelt jiggling. She tossed the blond bangs out of her eyes and (though she wasn't supposed to) reared up to put her front paws on Scott's arms, trying to give Connor a long lick. "Casey, move!" Scott groaned, tripping over her on the way to Connor's bedroom. Casey turned into a silly carefree puppy on Connor's homecoming days. She consented to dash outside to pee, then flew back through the door and galloped down the hall to leap onto Connor's bed before the Millards could waylay her. Connor stirred and mumbled, "Ay-ee," and she wagged and panted and looked up at Deb and Scott with twinkling eyes as if to relay: *You see? He's fine! Just give him a minute!*

DO DOGS—LIKE HUMANS AND OTHER PRIMATES—display "prosocial behavior"—a willingness to help another individual without deriving personal benefit?

Recently working with sixteen dogs on an experiment modified from similar ones with primates, Dr. Friederike Range and a research team at the Messerli Research Institute in Vienna set up a pair of side-by-side cages, introduced a dog into each cage, and offered the subject dog a choice between two ropes: with one rope, the dog could drag an empty tray to the cage of the adjacent dog; with the other rope, the subject dog could pull over a tray containing a treat to the adjacent dog. "The only purpose of the task was to benefit the other dog. By conducting several control tests, the researchers excluded the possibility that the dogs were simply pulling the trays for the fun of it." (All the subject dogs got a chance to choose a tray for themselves at the end of the test and all gladly hauled over the tray with the treat.)

Did dogs seem to notice or care whether another dog got a treat? They did! And if the dog was a friend, the subject dogs pulled over treats much more frequently. They did the work with the sole goal of benefiting another dog. "Dogs truly behave prosocially toward other dogs," the researchers concluded. "That had never been experimentally demonstrated before. What we also found was that the degree of familiarity among the dogs further influenced this behavior. Prosocial behavior was exhibited less frequently toward unfamiliar dogs than toward familiar ones."

"It is hard to believe that about ten years ago scientists concluded that only humans care about the well being of others," Frans de Waal commented on these results. "They were convinced that animals don't. Then came the chimpanzees, the monkeys, the bonobos, the rats, and now the dogs, to tell us otherwise."

Though it can't be experimentally tested, it does appear that many of Casey's actions toward Connor were "prosocial": performed not for himself, but for Connor.

Do dogs empathize?

Empathy—the ability to understand another's feelings and to enter in and share as much of the pleasure or the burden as possible—has been perceived by 65 percent of polled dog owners. That surprises me. I'd have thought Empathy would score 100 percent. Empathy is perhaps what dogs do best, as anyone knows who has tried to recover from surgery, illness, dashed hopes, or a broken heart with a dog nearby. One can't even roll over in bed when sick or depressed because the dog has glued itself to one's body. One can't fall asleep because the dog, now lying on one's chest, relentlessly attempts to make worried eye contact.

Casey's empathy for Connor was on touching display throughout the months of Connor's decline; when Connor felt a little lively, happy Casey rejoiced; but when it was a sick day, Casey lay quietly beside Connor on the sofa, intermittently offering a raised eyebrow or gentle wag of the tail.

Some skeptics believe that what passes for Empathy in dogs and other animals is actually a more primitive response known as "emotional contagion," in which one individual reflects another's emotional state without grasping the reason. Emotional contagion occurs in hospital nurseries when one baby starts to cry and soon they're all bawling. From this angle, when a dog sees a human's emotional distress, the dog feels distressed, too, and goes to its human not to extend comfort but to seek it.

And there are other researchers who believe that dogs who nuzzle a person in distress do so not out of discomfort or unhappiness, but only with curiosity (which suggests that they *do* give dogs credit for the complex emotion of Curiosity).

Two British psychologists designed an experiment to study "empathic-like responding by domestic dogs to distress in humans." Deborah Custance and Jennifer Mayer of Goldsmiths College, London, tested eighteen dogs in the dogs' homes. Two people in each residence—the dog's owner and a stranger—either hummed in an odd staccato way or pretended to cry. The theory was that if dogs are motivated primarily by *Curiosity* in response to crying, then they will approach and behave similarly in response to strange humming. And if "emotional contagion" is at work when humans cry, dogs should seek comfort from their owners both when owners weep *and* when strangers weep. (I'll go ahead and state the obvious: like the experiments in which humans feigned heart attacks or pretended to be pinned under falling bookshelves to fool their dogs, an experiment relying upon humans tricking their dogs by "crying crocodile tears" rests on a shaky foundation.)

At any rate, the dogs came through with flying colors.

When the humans hummed weirdly, dogs looked up, but did not approach them. Curiosity was satisfied from across the room.

When owners cried, their dogs approached and offered comfort in the usual furry sloppy wet endearing ways.

And when *strangers* cried, dogs approached the stranger and offered the same sweet messy condolences.

Dr. Stanley Coren, of the University of British Columbia, leans entirely toward Empathy in his summary of these results: "In the same manner that young humans show empathy and understanding of the emotions of others, so do dogs. Furthermore, we appear to have bred our dogs so that they not only show empathy, but also show *sympathy*, which is a desire to comfort others who might be in emotional distress . . . Your dog really does care if you are unhappy."

DID CASEY PERCEIVE THAT CONNOR returned home from every hospital visit enfeebled, diminished, miserable? The Millards didn't know what Casey detected or understood. She'd always been gentle with him. Casey and Connor had never wrestled or played tug-of-war. The roughhousing they did was all on Casey's side, almost without physical contact. It included the grunts and blurry outlines and flying fur of boyish tussling, but it was Casey mock-growling and gnawing on a toy or a sock, shaking her head as if to extinguish its life, or going through the motor patterns of chase and capture and kill-bite with the tennis ball that Connor feebly allowed to roll out of his hand. There really were no behaviors she needed to scale down or modify now that Connor was so fragile; she already treated him with unthinkable gentleness, while making him feel adventurous. All she seemed to ask of Connor now was not physical play, but just a playful mood, an open eye, a wry half-smile, a raised eyebrow. If he'd been able to sit on the chaise longue on the deck and watch her, she'd have frolicked her heart out, she'd have chased her tail, she'd have done anything to make him laugh. His physical weakness didn't stymie her. She patiently waited for the inner boy—the little kid who was always in there—to smilingly return and yell, "Ay-ee!"

But long days passed without the little kid putting in much of an appearance. The cautious trip, in his mother's or father's arms, from his bed to the den sofa with the attached oxygen equipment cart became each day's most ambitious outing. Casey did what she could with it. She pranced ahead of Deb or Scott, tail aloft, head held high, a drum major. She did some sideways leaps in the den as they got Connor settled on the couch and turned on PBS children's programming. At the sound of a theme song—*Sesame Street, Barney,* or *Curious George*—she splayed her front feet out in front of her and lifted her rear end, ready to rumble. These songs had been the soundtrack of her happy life in this house. If Connor lay motionless on the sofa, Casey hopped up and reclined optimistically just beyond his feet. She waited. After an hour or two, if he awoke a little, she sat up, wagging. When he stretched his thin white neck feebly toward the TV or just painfully closed his eyes again, she watched his face closely for long moments, alert for a signal. Her floppy ears lifted slightly so as not to miss a whispered cue. Then she exhaled and lay down again. "Casey, outside? Walk?" Deb called hourly, but the dog wouldn't leave Connor unless Deb walked over, clicked on a leash, and insisted.

Sometimes, at the far end of the sofa, Casey sat up, looked around, and evidently concluded that she could be closer to Connor. She squirmed along his length and stretched out with her head on his pillow. Sensing her, Connor moaned a small welcome, rolled onto her, or lifted his arm over her back. Casey audibly sighed. This would do.

ONE DAY IN LATE 2006, Connor's fever hit 105 and he fell unresponsive. In the hospital, the doctors found an infection in his bloodstream, a sepsis. To everyone's surprise, they discovered the presence of an extremely rare blood disorder, hemophagocytic lymphohistiocytosis (HLH). "He has this freakish thing, *too*?" cried Scott. HLH was an "orphan disease," affecting only one in a million. "Don't look it up," a doctor suggested. "It's bad."

At home, online, the Millards looked it up immediately and learned that HLH destroyed bone marrow. The only possible response was a bone marrow transplant, which Connor couldn't survive. The Millards truly grasped now that their son's life would be short, that he was running out of time. At home, he fell victim to countless infections; his blood pressure was erratic; he ran high fevers; he was sore, listless, and disoriented. When they wrapped him in a light blanket and ran to the car for another trip to the emergency room, he floated in their arms like a baby.

In September 2007 they reached the end. Almost paralyzed with shock and dread, they brought Connor home from the hospital for the last time. For the last time, as they came through the front door, Casey raised her head high and marched around the living room in jubilation, brandishing her tail like a banner. She leapt forward and splayed her front legs in her merriest play crouch, but they were already shuffling toward Connor's bedroom with her wisp of a boy and his equipment cart.

DO ANIMALS GRIEVE?

It is said that grief is the price we pay for love; that, where there is no love, there is no grief.

So, do animals love?

Do *dogs* love?

Do they love *us*?

We certainly love dogs, and we have the data to prove it: the way we live healthier and longer when we live with dogs, the ways we feel more contented and happy and less alone and less anxious when we spend time with our dogs.

But do dogs love humans? Putting aside the world's free-ranging, homeless, feral, abandoned, or mistreated dogs, how does the average house-dog feel about its resident human?

The experiments detecting the spike in "feel-good" hormones in humans petting their dogs found that the same "happy hormones" spiked in the dogs, too, when their humans stroked them or gazed into their eyes. It relaxes them to be near us, as it relaxes us to be near them.

Is there anything more? Would they give us cute names, buy us little raincoats, and write poems about us, if they could?

In Atlanta, in 2010, a scrunchy-faced snuffly little pug named Newton died at fifteen years of age. His grief-stricken owner, who happened to be an Emory University neuroscientist, found himself wondering if Newton had loved him as much as he'd loved the little dog. Unlike most people, Dr. Gregory Berns, distinguished professor of neuroeconomics, had access to tools with which to probe for an answer. If dogs could be trained to lie still in an MRI scanner, he thought—awake and responsive rather than sedated—it might be possible to view their brain activity and to collect "neurobiological evidence of emotions."

He enlisted the help of a professional dog trainer and began the almost-thankless task of training dogs—his own and those of local volunteers—to lie very, very, very, very still in a noisy tube.

With the MRI-trained dogs, the research group was able

to zoom in on the active brains of quietly alert animals. They focused on each dog's caudate nucleus, a small region of the brain associated, in humans, with the "reward system." "Rich in dopamine receptors, the caudate sits between the brainstem and the cortex . . . [and] plays a key role in the anticipation of things we enjoy, like food, love, and money," Dr. Berns writes. "Caudate activation is so consistent that under the right circumstances, it can predict our preferences for food, music, and even beauty."

The researchers first established that the canine caudate lit up in response to anticipated pleasure, just like the caudate in the human brain. When a researcher displayed a neutral hand-signal to a trained dog, nothing happened; but when a researcher flashed the agreed-upon signal for "hotdog," the animal's caudate lit up just like the human caudate does when anticipating pleasure.

Then they designed an experiment in which five odor samples would be presented to a dozen MRI-trained dogs. Which scent would generate the greatest sense of anticipated pleasure, that of (1) a dog unknown to the dog in the scanner, (2) a dog friend, (3) a human unknown to the dog, (4) its own human, or (5) the subject dog itself?

While the olfactory bulb was activated every time, "the caudate was activated maximally to the familiar human." In other words, at least for these twelve dogs, the whiff of their own human gave them the greatest sense of anticipated pleasure. For Dr. Berns, it is evidence of "a natural interspecies bond that is stronger than the dogs' innate intra-species bond . . . [and of] the importance of humans in dogs' lives."

In a later test, the research group asked each dog's owner to step out of view momentarily, and then to return. And there it was again: the dog's pleasure center lighting up at the reunion with the owner.

"Do these findings prove that dogs love us?" writes Dr. Berns. "Not quite. But many of the same things that activate the human caudate, which are associated with positive emotions, also activate the dog caudate . . . and it may be an indication of canine emotions. The ability to experience positive emotions, like love and attachment, would mean that dogs have a level of sentience comparable to that of a human child."

In the absence of their buying us rhinestone-studded collars and squeaky squirrels, fMRI images of dogs' happy thoughts about their humans will have to suffice, for now, as proof of their love.

IN THE MODEST HOUSE IN Mt. Arlington, the next set of days was like time spent in a room without windows, not marked off by bedtimes or mornings, sunlight or sunset, lunches or dinners or breakfasts. Dirty plates bloomed in strange places. Scott didn't leave for work. No one slept in the normal spots—Connor stayed on the den sofa, Scott sprawled out on the floor beside him, and Deb collapsed on Connor's bed, cradling his football pillow. Pediatric hospice helpers quietly came in the front door, spoke in low tones, and squatted beside Connor where he lay on the sofa. People forgot to feed Casey, or poured food into her dish at odd hours, like 3:00 a.m. Here and there across the house, lights stayed on all night. Clock radios buzzed into action, blasting news at 6:55 a.m. as if from a foreign time zone. Casey got taken outside after midnight; she and Scott walked down the street in darkness, under the hilltop dazzle of stars. When had Casey last seen Connor's crooked endearing smile? Connor's trains were a mess, the little cars and engines and tracks shoved against the wall of the den. Even his inseparable Casey-the-stuffed-dog had fallen behind the stack of children's videos and DVDs, gathering dust. In the backyard by herself, Casey discovered an old tennis ball in the grass. She raised her head and mouthed it for a happy moment, then dropped it, looked at it, and left it alone. Groaning, she dropped to her belly outside the kitchen door to await reentry. She whimpered aloud occasionally, to remind them. A nurse opened the door. Scott took Casey by the collar and didn't allow her to spring over to Connor. She veered sharply left, within gazing distance of the pale still child on the sofa beside his mother, but she couldn't get close enough to give him a lick or to detect any lip movement or murmur of "Ay-ee." Scott pulled her down the hall to Connor's bedroom, ruffled her hair, said firmly, "Stay, Casey," gave a single sob, and closed the door.

An hour later, released from the bedroom, Casey reached Connor's side in a few rollicking leaps, but she came to an abrupt halt and inhaled the news that Connor wasn't there, or that what was there wasn't Connor. "Oh, Casey!" cried Deb, awash in tears, her straight hair slicked back, her voice swollen, her face underwater. Casey turned away and walked with dignity to the living room, straight to her bed. She sat on it and watched the front door. In the following hours and days, people came and left through the front door; Deb and Scott left and then returned and not-Connor went out with strangers and did not come back. Flowers came through the front door; hot casseroles came through the front door; dressed-up soft-spoken people came through the front door; Connor's grandparents came over to Casey's corner and stroked her head and tickled her chin and Casey licked her lips and blinked her eyes in docile appreciation. Finally everyone left except Deb and Scott, who resumed the non-schedule of shapeless non-days and non-nights, front door ajar in the night, lights staying on in the afternoon, plates of food sitting out on the kitchen counters. Instead of sleeping, they wept. Instead of eating, they sobbed. Instead of getting dressed, they wandered around confused, barefoot, uncombed, in T-shirts and pajama bottoms. They pointlessly moved a dirty plate from the living room coffee table to the bathroom sink. They pulled towels and clothing from banks of dirty laundry and used them. Casey sat silently on her bed, erect, watching the front door. Occasionally, from need, she stood silently at the back door, relieved herself in the grass, returned to stand expressionless at the door, and then walked back to her bed.

After some period of shapeless time, Scott returned to work, Deb returned to bed, and Casey circled. She circumnavigated Connor's bedroom, drinking in the fading scent of him, then she visited the front door—*in case this was a homecoming day!*— and then she returned to Connor's bedroom. She nosed around Connor's laundry hamper and bed, snuffling up stray molecules of skin and oils and lotions. Then she completed the lap back to the front door—*in case he was coming home today!* Like a caged zoo animal, she paced the lopsided oval ten, twenty, thirty times until Deb couldn't bear it anymore. "Casey, stop it! No! Casey, stop!" Casey stopped. She walked to her bed in the corner of the living room, tried resting, didn't feel tired, sat up, watched the front door—*in case!*—then walked back to Connor's bedroom. She snorted at the crack under his closet door and then inhaled, drawing Connor's smell from his clothes down and out through the crack at the bottom of the door. She sniffed the boxes of backup equipment, of hoses and trachs, without much interest, finding only Deb and Scott there and whoever had packed the boxes and whoever had delivered them, and she peeked into the bathroom, drawing in molecules of odorants from Connor's shampoo, Pull-Ups, baby powder, and toothpaste; and she walked down the hall, head lowered, seeking whiffs of him; and she reached the front door, *in case*. Really unable to bear it, Deb closed herself in her bedroom and the house stood silent other than the whisk and flick of paws and toenails up and down the hall, back and forth, back and forth, back and forth.

DO DOGS GRIEVE?

"Almost no scientific research has been carried out on dog grief," says Dr. Barbara J. King, professor of anthropology at the College of William and Mary. "A recent wave of studies into aspects of dog cognition, however, supports the notion that dogs are incredibly sensitive to others around them." The American Society for the Prevention of Cruelty to Animals conducted a Companion Animal Mourning Project in 1996. The study found that "36 percent of dogs ate less than usual after the death of another canine companion. About 11 percent actually stopped eating completely. About 63 percent of dogs vocalized more than normal or became more quiet. Study respondents indicated that surviving dogs changed the quantity and location of sleep. More than half the surviving pets became more affectionate and clingy with their caregivers. Overall, the study revealed that 66 percent of dogs exhibited four or more behavioral changes after losing a pet companion."

Countless dog and cat owners have witnessed (as I have) the longing displayed by an animal who has lost a loved one. I grew up with a bonded pair of blue-eyed seal-point Siamese cat sisters—identical except that Dusty had crossed eyes, a kink in her tail from an early encounter with a refrigerator door, and an odd anxious habit of muttering in vexation whenever the vacuum cleaner emerged from the closet. Smokey, the real beauty, lived to be eighteen years of age; crotchety Dusty survived her by a year, but in confusion and sorrow and diminished appetite.

The guttural yowls with which she called again and again, over and over, up and down the halls, day and night, in her relentless search for her sister brought us to tears.

Dr. Coren witnessed similar bereavement when his dog Dancer, a Nova Scotia duck-tolling retriever, outlived his lifelong best friend, Odin, a flat-coated retriever. "With Odin now gone, Dancer systematically looked at each of the four locations where his friend would go to lie down . . . It was several weeks before he stopped checking all of the places that Odin should have been . . . Much like one might expect from a child who did not have the concept of the permanence of death, Dancer never gave up on the idea that Odin might reappear. Up through the last year of his long life, Dancer would still rush toward any long-haired black dog that he saw, with his tail batting and hopeful barks as if he expected that perhaps his friend had returned."

Though I have not personally witnessed displays of desolate searching among the dogs in our household, I have often fantasized the leaps and yelps of joy that would ensue if a certain beloved deceased dog (or young man) were suddenly to return to our yard—as if the absence had been temporary and the reports of death greatly exaggerated, and now dogs and boys alike could get back to the business of running in circles, wrestling, and laughing.

According to Dr. Berns, fMRI scans of dog brains could not confirm or deny the existence of grief because "it's unknown how grief looks in the human brain . . . If I were to speculate, I would guess that, like people, some dogs mourn and others don't."

"It's bad biology to argue against the existence of animal emotions," writes Dr. Bekoff. "Emotions have evolved as adaptations in numerous species, and they serve as a social glue to bond animals with one another . . . Emotions, empathy, and knowing right from wrong are keys to survival, without which animals—both human and nonhuman—would perish. That's how important they are . . . Emotions are the gifts of our ancestors. We have them and so do other animals. We must never forget this."

"To endow animals with human emotions has long been a scientific taboo," writes Dr. de Waal, "but if we do not, we risk missing something fundamental, about both animals and us."

"SCOTT AND I LOST A CHILD, but you and I lost our jobs, too, didn't we?" said Deb to Casey one afternoon, drawing the dog out of her infinite ovals. They sat together on the den sofa. Deb gazed into the backyard and stroked Casey's broad back. "You were amazing, Casey, you know that? Did I ever tell you how amazing you were? You took such pride in your work. It was honorable work and you did it with such drive and with so much love. Your love showed in everything you did. And he loved you, too, so much." Something here echoed in Deb's mind . . . oh, Connor sitting beside Casey in this very spot—after refusing to bestow a single sentence on the poor speech therapist—prattling on and on, chirping like a little bird, telling only Casey his stories and secrets. "You loved Connor so much, Casey, I know you did. You don't know what to do with yourself now. And neither do I."

CHAPTER 15

Logan & Juke

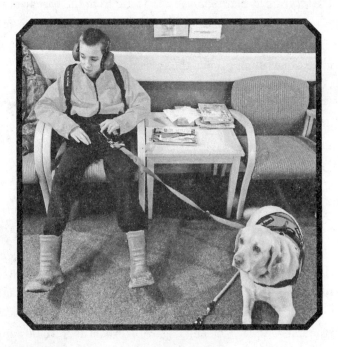

J eff Erickson and Iditarod musher DeeDee Jonrowe met
in the 1970s when he was a deep-sea fisherman and she
worked for a major fish-processing company. After both
married, the two couples became good friends. In 2009, Unal-

akleet School hosted a regional high school wrestling and cheer-
leading tournament and Jeff asked DeeDee to serve as a celebrity
visitor, to inspire the young athletes. She arrived by plane, rather
than by dogsled, and accompanied by one dog rather than a dozen.
In the Iditarod off-season, it was Mr. Myagi the Pekingese's turn
to see a bit of the world. By the light of the wood-burning stove,
DeeDee sat with her friends, enjoying salmonberry pie and hot
tundra tea.

"He runs all night, turning on all the lights, radios, TV,"
Donna told her. "We sleep in shifts. It's not called 'sleep.' "

"Everything's blaring noise and static from the basement to
the bedrooms," Jeff said. "The other day, I turned down *one*
radio in the basement when he was two floors up in his bedroom,
and he came running down the stairs to see what happened."

"We're wiped out," said Donna.

"If you get five hours of sleep, you're like, 'Wow, five hours,' "
Jeff said.

"It's a living hell," said Donna.

Jeff left the house and walked up to the high school, where he
was spending the night as a chaperone.

Alone with her old friend, Donna said: "To tell you the truth,
I feel like autism is tearing us apart."

It was a blustery night, the dark wind raging and pushing
against the house. High wind and storms often pitched Logan
into freak-outs of a scale only Jeff could subdue, and Donna's
teary recitation was frequently interrupted by nervous glances
up the stairs. At any moment, Logan's bedroom door would bang
open and he'd emerge screeching and clawing at himself and rip-
ping off his pajamas and diaper.

As the wind came shrieking off the sea, Logan's bedroom
door did open. To Donna's surprise, he walked downstairs with

a look of curiosity, spotted Mr. Myagi, who was curled beside DeeDee on the sofa, sat on the floor, and began stroking the little dog, seemingly oblivious to the whirlwind outside.

Then the night got crazy. An alarm sounded that the waters were rising and houses along the beach must be evacuated. Outside, DeeDee tucked Mr. Myagi inside her coat and Donna clung to Logan and the three of them were tied into a rope-line with neighbors as the wind reached speeds of 80 miles per hour. They bent over and struggled to walk the few blocks inland. They were guided to a house where dozens of people waited out the storm.

Donna accepted a small bedroom for herself, DeeDee, and Logan. *Any minute now, Logan will flip out and keep me and everyone here up all night,* she thought. *These folks have never heard anything like the explosion that's about to happen.* DeeDee released Mr. Miyagi onto the double bed. As storm winds pounded the village, Logan focused on the little dog, petting him until both fell asleep.

The next morning, after the storm, as DeeDee and Donna picked their way through the debris back to the Ericksons' house, they noticed Logan was moving his lips, making a sound. Logan hadn't said a word since his cognitive collapse at age two. They bent in to listen and were able to make out Logan's word: "Myagi."

DeeDee said: "Donna. We have to get this boy a dog."

SHE SAID GOODBYE to the Ericksons and flew east toward her home in Willow and the slightly more inhabited parts of the state. Donna accepted DeeDee's suggestion that a dog could do Logan some good, but she had no energy to pursue it. She had no energy even to long for a life different from this one, which she was trying to survive hour by hour. DeeDee Jonrowe did have the energy, however, as well as the curiosity and the optimism to investigate. She was an expert on breeding, and it wasn't hard for her to imagine that someone, somewhere, bred dogs to assist kids like Logan Erickson.

A LITTLE BOY NAMED LEO BERNERT lived in Anchorage. The youngest child in a close-knit family of blond, suburban, winter sports–loving people, his development, like Logan's, had been stymied by autism. "He had a normal infancy . . . until he wasn't all that normal," his mom, Peg Bernert, told me. A gracious, easy-going woman—practical, and a good sport—she looks like a California blonde on a ski vacation in her lined winter tracksuit and shiny snow boots. The motherhood arc she was merrily tracing—from PTA opening meeting to Halloween to Thanksgiving to Christmas, with the sprinkling of birthday parties in between—sputtered to a halt when Leo's development nose-dived.

"Leo as a baby didn't demand attention from people, didn't look around," she says. "I found myself saying to people—to kids—'Say hi to Leo,' which I'd never said with our other children. He cried a lot. He had extreme separation anxiety—I couldn't leave the room. His speech was delayed.

"For my birthday the year Leo turned two, I flew to Seattle to attend the Notre Dame (my alma mater)/University of Washington game with my sister. That night before bed I was just browsing my sister's bookshelves and saw a book called *The 10 Signs of Autism*. I sat up all night crying over that book. Leo had eight of the signs—you know, crying a lot, twisting his fingers, lining up his little cars . . .

"We took him to a three-day appointment with a pediatric neurologist in Anchorage. The third day, he handed down the verdict. He was a really kind, good man. He said, 'Listen, at this point in time you have no idea where this child is going to go, he has all the potential in the world, and we're not going to place him. When he's an adolescent, we'll have a better sense of where his life is going to go but right now it's wide open. Don't compromise the dreams of your family because of this. Do not make him the focal point of your family.'

"I cried all the way *to* the meeting, I cried *through* the whole meeting, and I cried the whole way home. At the meeting, the nurse kept asking me if I had a cold. I thought, *Don't you know what's in store for us? I've educated myself and I* do not *want to be here.*

"I'd only known one kid before then with autism. He just had odd behaviors: he'd go into people's houses and turn all the lights off and on. Wasn't real focused. I was kind of fascinated by him. He's Sarah Palin's nephew."

LIKE SO MANY CHILDREN on the autism spectrum, Leo was an eloper.

One winter, when the family lived at the edge of a national forest, Leo disappeared at night, barefoot and in his pajamas. An increasingly frantic search through the woods and along a stream turned up nothing—and he didn't answer to the sound of his name. "I was running down the freezing road with the feeling, *He's gone,*" Peg says. "He liked to throw pebbles into a culvert and my husband, Dave, and I were afraid he'd gone there and fallen in. I was way down the street, running, on the phone with 911, when our daughter Mackenzie came running toward me, holding Leo." He'd been discovered shivering violently, crouched behind cardboard boxes on the cement floor of his own garage. Holding him close, Peg thought: *A dog would have found you.* Within the year, the Bernerts had discovered Karen Shirk and applied to 4 Paws for Ability, the first family from Alaska to contact them. In 2007, the Bernerts attended the ten-day workshop in Ohio and brought home a golden retriever named Halo, trained in autism assistance and tracking.

"The first time Halo tracked Leo, I felt like singing from the mountaintops," she says. "We saw that the door had been left open a crack and said, 'Track, Halo. Where's your boy? Where's your boy?' She was deeply asleep and she jumped up and ran out the door. She found him way down the street. After we got Halo, everything changed. Leo realized we always found him. The magic went out of running away. Now he doesn't even bother anymore."

Though she wasn't expecting this side effect, Peg found her lonely life with Leo dramatically changed, too. "My son is really beautiful," she says. "He has ash-blond hair, ice-blue eyes. He looks just like Dave, who's a great outdoorsman, a fisherman and skier. But our son is just . . . vacant. We say Leo lives in his own zip code. One day in the kitchen, I was facing him, talking to him. I said, 'I love you so much, but you're just not here with me, are you?' I let him go and felt a tremendous wave of sadness. Then I felt a pressure on my shoulder. I turned and looked straight into the eyes of Halo. She'd come up behind me, and then stood up. It was to comfort me. It was unquestionably an act of empathy."

Peg, who'd always loved dogs, returned to 4 Paws for additional training, began fostering puppies, and then completed a training internship under Jeremy. Today she is the head trainer for 4 Paws Alaska, in Anchorage.

"Look around the room," she told me one day in the training circle in Xenia. "A lot of these dogs, though trained as autism assistance dogs, will become therapy dogs for the parents, too, especially the moms. The moms don't know it yet, but Karen and Jeremy know it. They know how badly these mothers need a friend."

PEG BERNERT RESPONDED WARMLY WHEN contacted by the famous DeeDee Jonrowe about the Ericksons. At the next meeting of the Iditarod Trail Committee, DeeDee and Kathy Fiedler, the wife of a musher, made a fund-raising pitch: "We have the charity for you this year. Let's buy a service dog for an Iñupiaq kid in Unalakleet." The committee agreed that the 2010 race fundraiser should be called "A Service Dog for Logan." World-class mushers began pledging portions of their winnings to pay for a 4 Paws autism assistance dog.

Not until then did DeeDee tell the Ericksons the plan that was already in motion. "Don't worry about a thing," she told them. "We've got this. All you need to do is make a list of the things you'd like a dog to do for Logan."

"You won't believe this, but a few years ago I clipped an article from the *Anchorage Daily News* about Peg and Leo and their dog," Donna replied. "I probably still have it around here somewhere."

"With DeeDee and Kathy having already made the decision, it was easy for us," Jeff told me. "The *decision* was the big thing. We had so many people supporting us, so many of the mushers . . . all we had to do was jump on board because the train was leaving the station. We phoned Peg and Dave for information. They got us excited; they gave us a vision."

They compared the moments of diagnoses. "I came home horribly devastated," Peg told them. "I did not want this path. No. No. No. Some people are like, 'I'm going to read all the books about this,' but I didn't want to read them. I think some people can just place their energy—all this new energy, about this new thing that's going to affect the rest of your life—some people just dive right in and get started. That was not me."

"Me either," said Donna.

"The first time we got a full night's sleep was at the hotel in Xenia on our first night with Halo," Peg said. "I woke up completely astonished. But I thought, *No way. This won't last. It's just the excitement of the big day. Leo is worn out.* But it did last. The presence of Halo helps Leo sleep through the night. Listen, we'd have paid good money if Halo had done *nothing* for us other than let Dave and me sleep through the night. I call 4 Paws dogs 'Sleep-Through-the-Night Dogs.' "

NINETEEN TEAMS RACED FOR LOGAN in the 2010 Iditarod. At the Unalakleet checkpoint, Jeff and Donna, with Logan, greeted as many as possible when they pulled in, with gifts of dog booties and smoked salmon dog treats. The grizzled men and wind-burned women saw to their dogs, stood up to get their bearings and stretch their backs, and saw Logan. "Hey, buddy," they said, kneeling down to greet the silent, uncomfortable nine-year-old. Many were world champions: Martin Buser, Aliy Zirkle, Jeff King, Aaron Burmeister, Mitch Seavey, Dallas Seavey, and of course DeeDee Jonrowe. "We're going to get you a dog, you know that? You're going to have your very own dog." "Next year this time I want to meet your new dog, okay, kiddo?"

Jeff says, "The front-runners involved in the greatest sled race on earth pulled in here and came up to Logan and got on their knees and started crying. Lance Mackey said, 'You'll never regret it, Logan, you'll never regret it. Having a dog is going to help you so much.' Logan was impassive as usual, while this great musher was crying. You know he went on to win the race? He won the Iditarod and he was racing for Logan."

IN SEPTEMBER 2010, AFTER TWO weeks in Xenia, the Ericksons brought home Juke, a purebred yellow Lab. Bulky and calm, he quietly accepted his place in the family and got down to the business of focusing on Logan. Though he was still a young fellow, he was a very earnest animal, with an impressive sense of duty. From the first, Juke asked little for himself. Some dogs whine and bang about and run from window to window and glance up frequently at the faces of their humans to express their intense desire to go outside and bark at stuff. Juke did not. Donna and Jeff wanted him to be happy. He was so restrained and responsible that, at first, it was like having a very polite foreign exchange student in the house, who they hoped would feel at home. When they whistled and called up the stairs to him from the front door, wanting him to go outside and enjoy a few minutes of dog life, he came down the stairs compliantly and stepped out to use the bathroom and sniff the air, but he certainly didn't bowl them over on his way out.

He made friends, though. Once outside in the bracing air, through which the barks of dogs clanged constantly, he frolicked a bit, dancing eight or ten steps when three or four might have covered the distance. Nose down, he carved a route through the packed snow. He circled behind the house and crossed a few neighboring backyards; he greeted dogs on chains and dogs that were loose. Having grown up at 4 Paws, he was a sociable guy, knew the etiquette, the postures, the sniffs, and the permissions-to-sniff.

With human beings, he made friendly eye contact. He sat on command. In this dog-loving village, he quickly became a beloved dog. "Good morning, Juke," pedestrians said when they saw him trotting beside Logan and Donna on the way to or from school.

At the 2011 Iditarod, Donna, Jeff, Logan, and *Juke* made their way down the icy slope to stand by the frozen river and wave in the sled teams. The mushers, who'd been too moved to say much the previous year, now could barely summon any words at all as they knelt to look at Logan, then at Juke, stroking with thickly gloved hands the boy's down jacket sleeve, and the broad back of the dog. "Logan," they croaked. "This your dog? What's his name? Good boy, Juke, good boy, you're a good boy." Shaking their heads, weeping, they limped up the slope toward hot coffee, scrambled eggs, and a nap.

Inside the Ericksons' house, a new style of life began. It included sleep. It included an early warning system if a great meltdown lay ahead. It included not fearing that Logan had let himself out of the house, slipped into the sea, and drowned. "When Logan 'escapes' now, he does it like a game," says Donna. "He knows he can't outrun Juke. When we're out by the water, Logan will run down the beach like he wants to get away, but he keeps looking back to make sure Juke is after him. Juke overtakes him and gently herds him back toward the family."

"In the morning, when they wake up in Logan's bed, if Logan is touching Juke in any way, Juke won't budge," says Jeff. "This morning I told him to go potty and come eat and he basically scowled at me and shook his head, like, *I'm working.* Logan was asleep but his arm was touching Juke, so Juke refused to leave him."

In March 2013, Karen Shirk visited Unalakleet and got a glimpse of Iditarod dog teams skimming toward town, or away, on the Unalakleet River. One night, she went with Donna to a church pie supper. The pride of village kitchens, presented in Tupperware dishes, included berry harvest pie; Eskimo ice cream made with reindeer tallow; *Akutaq* made from seal oil, fish row, and berries; and *Muktuk,* the frozen whale-skin treat. Other traditional desserts had ingredients that included whipped fat, meat, roots, and leaves. Karen gamely filled her paper plate and tried a taste of everything. At their paper-covered folding table, Donna introduced Karen to her friends and word quickly spread. Now the room's center of gravity shifted away from the delicacies and toward Karen Shirk. People of all ages began drifting over to talk to her and shake her hand or give her a hug. From a long way off, moving very very slowly, came the village's oldest people, octogenarians and nonagenarians, tiny ladies with faces wreathed in smiling wrinkles, wrapped up in fur-cuffed parkas and scuffing in fur-lined boots, as well as ancient bronze-hued men, toothless and weathered. They crossed the room in pilgrimage to Karen, almost standing in line to greet her. "Thank you," they told her in heartfelt tones, taking her hands in theirs. One little lady bent close and whispered to Karen: "I am very thankful for Juke." An elderly man said with sonorous formality: "Thank you. Thank you for what you have done for our village."

IN 1910, CAPTAIN ROBERT FALCON SCOTT, a British naval officer and explorer, set out to discover the South Pole and claim it for the British Empire. He purchased Manchurian ponies for his expedition, against all warnings that this was a dire mistake. The greatest polar explorer of the day, Norwegian scientist, diplomat, and Nobel Peace Prize laureate Fridtjof Nansen (who had led the first crossing of Greenland and had come the nearest to the North Pole of any nineteenth-century explorer), offered these words to Captain Scott: "Dogs, dogs, and more dogs." But Scott had been unsuccessful on an earlier expedition with dogs, and turned a deaf ear. The British team set off across the continent of snow and ice with ponies.

Simultaneously, a Norwegian explorer, Roald Amundsen, and his team landed and sledded across the ice toward the South Pole from another direction. They hoped to arrive first and claim it for Norway.

Unlike the Brits, Amundsen's team traveled with dogs and made good time.

While the Brits hauled feed for the ponies, the Norwegians killed penguins and seals upon which their dogs feasted. The Norwegians knew how long to rest the dogs (up to sixteen hours a day), how much to feed them (each dog required the same quantity of meat as a man), and how to keep up with the dogs and not hold them back (wear skis).

Captain Scott's ponies were in no way equal to the task. They all perished, and were eaten, after which the British explorers dragged their own supplies. When they finally staggered into view of the South Pole, they beheld a nicely furnished tent, crisply flying the flag of Norway. Inside was a friendly note to Captain Scott from Amundsen. He had arrived a month earlier, claimed the South Pole for Norway, and ridden away with his teams of sled dogs.

Like the South Pole, there are uncharted and perilous reaches of the human psyche accessible only when accompanied by dogs.

CHAPTER 16

Eddie Hill & Timber & Dante & Keeper & Jiminy

Twelve years into Eddie Hill's sixty-year sentence, a local animal shelter showed up at WCI and invited volunteers to foster and train hard-to-place dogs. Eddie applied. He was stunned and overjoyed when his application was accepted. Eddie Hill was offered Timber, a German shepherd mix suffering from severe neglect and malnutrition.

Timber was no great prize in the outside world—if he didn't succeed in the prison class and was deemed unadoptable, he'd be euthanized—but to Eddie, he was a miraculous being. It was as if a mythological beast had suddenly materialized in his cell. "Someone had put a collar on him when he was a puppy and left it there, never widening it as he grew," Eddie told me. "His skin and fur grew over it and had to be cut away; his neck was a mess. He was terrified of everyone and everything—the loud echoes, yelling and banging, people and dogs, and wheeled carts. He constantly cowered and put his tail between his legs and shook all over. I think he had never known kindness."

Eddie was awed by the animal's pain, and awed that he'd been entrusted to heal him. He understood that Timber was a throwaway, but then he was basically a throwaway himself.

On their first day together, seeing Timber's terror and despair, Eddie led him away to a quiet place on the grass outside. "I just kept petting him, looking at him, stroking him, and telling him over and over, 'It's all going to be better now, boy. You're safe now. From now on, your life is going to be great. No more bad people.' "

These were promises he couldn't make to himself, of course, but he was stirred to be able to offer these words to the beaten-down dog, an innocent creature who was more alone in the world and had known more suffering than Eddie had himself. "I attended training classes and started devouring every possible book I could find about dogs," Eddie said, "but a lot of what I did was trial-and-error, plus common sense and paying close attention. The first thing that struck me about Timber was: 'There's somebody in there.' All I had to do was look and he was *right there*, with his own feelings and fears and hopes. I felt like he needed me, and wanted to connect. I felt him trying so hard to be good for me."

Eddie and his cellmate—collaborating closely for the first time, out of mutual concern for the hurt dog—worked on Timber's raw, wounded neck, cleaning and disinfecting it daily. "He loved it, as it healed, when we would massage lotion into his neck to keep it soft. He was just in heaven."

Attuning himself to Timber's needs, movements, and dog language was a master class in dog training for Eddie. "It doesn't matter what kind of day you're having or what you're going through personally—you *have* to keep your cool, even with a sixty-pound dog jerking your arm in every direction and wrapping his leash around your legs trying to escape everything he hears or smells or sees as dangerous. To lose your patience is to go backward. The plan is for you to guide the dog *out* of the problems that humans created in the first place, not to create more problems for him."

But Eddie couldn't spare Timber the slamming of cells and racket of metal-on-metal; it's not like he could sling on a backpack, whistle to Timber, and disappear with his dog for a week of climbing in the misty Appalachians of his native West Virginia, though he longed to do so. And they were surrounded by inmates who thought it was funny to stomp their boots or clap their hands at the already-terrified animal, or to offer him food and then jerk it back, laughing when he dropped to the floor and cowered. There were limits to Eddie's ability to shield Timber. But with a warm voice, kind hand, and steady presence, he did what he could to help the dog weather everyday frights and minor cruelties. Little by little, Timber regained a tiny bit of courage. One day he briefly came out from behind Eddie's legs to sniff and be sniffed by another dog, before hurrying to hide again behind this man who was kind to him. On another day, he emerged to romp briefly with other dogs on the grass. Eddie and his cellmate gave each other high fives. Soon Timber was part of the prison pack, eager to play and chase. Eddie says, "In about seventy days, Timber went from a scared, hurt dog who didn't even know 'Sit,' to a pretty self-assured and clever fellow who could play with other dogs and do tricks and who was incredibly attached to me." The happiness Eddie felt in Timber's company was an emotional high that would have seemed impossible behind bars. It was one of the emotional highs of his life. The love he felt for the dog came pouring out of a heart that had been walled-in alone for a very long time.

Timber's rapid rehabilitation, however, meant that the dog might soon be returned to the shelter as a candidate for adoption. Eddie didn't know whether to put Timber's rising confidence and mastery of commands on display in the weekly class, or to try to camouflage his development. But it was out of his control. Timber pranced into the classroom a new fellow: tail up, head up, ears perked, eyes bright, coat smooth and glossy. He sniffed other dogs and let them sniff and never once tucked his tail between his legs or looked anxiously around for Eddie or ran back to hide under Eddie's chair. He knew Eddie was there, backing him up; he could do this. He was a dog among dogs.

Eddie was so proud of him, but his heart began pounding when the shelter staff member came over to pet and admire the new Timber. "Nice work. You think he's ready for a family?" Eddie had to agree that he was. "We'll pick him up next week." All week Eddie tried to block from his mind what was coming, and to impart to Timber every piece of love and guidance he would need for the rest of his life.

The following Tuesday, Eddie handed over Timber's leash, knelt, pressed his face against the dog's face for the last time, and stood back as his best friend was led away through the doors, gates, security checkpoints, and fences impassible to the man. Timber looked back to Eddie several times, then heeled obediently, as Eddie had taught him. He had no idea he would never see Eddie again. The man and dog had saved each other, but could not spend their lives together.

A FEW WEEKS LATER, in a small miracle of kindness, Timber's new owner wrote to WCI.

To Timber's Trainer:

I want to thank you for Timber. He is the most wonderful dog! Thank you for showing him how to be a gentleman! He is adapting to his new home very well. I am so appreciative for all your obedience training.

At night he lies down on the floor on a blanket and stays there all night until I get up in the morning and give him the "okay." It's amazing! I take him for a couple walks a day, which he enjoys. He wasn't sure about cars at first. He would stop in his tracks and watch them drive by . . . he's gotten much better. It's intriguing to watch him discover new things like butterflies and lightning bugs.

The other day he watched an American flag blowing in the breeze and just sat there and stared at it for the longest time . . . The playful puppy in him can be pretty funny. He sticks his butt up in the air, and runs and runs in circles . . . I give him plenty of time to be playful and then when it's time to stop or leave I call him to come and sit or stay and wow! he responds beautifully. THANK YOU SO MUCH . . . I show everybody what a wonderful job you've done with him. I think the training he received from you has made all the difference in the world for him . . .

I have 3 cats. My older cat and Timber are already friends. Timber licked her right upside her face the other day. The other two cats have been more shy but are warming up quickly. Timber gets so happy upon seeing them. He whines for them to come closer. I'm sure in time they'll better understand one another.

Timber is the best dog in the world. I'm so glad to have him. I hope you have the opportunity to train more dogs because you have done an awesome job with Timber. May God bless you.

Eddie stood in a prison foyer reading the letter, crying openly. He didn't care who saw him.

ON SEPTEMBER 20, 2012, AFTER twenty-two years on death row, forty-seven-year-old Donald "Duke" Palmer was executed by lethal injection for double homicide in Belmont County in 1989. A few months earlier, as his execution date approached, Duke had tried to undo one last bit of collateral damage. In an appeal to the Ohio State Parole Board, he wrote: "There is . . . another victim in this case—Eddie Hill, my so-called co-defendant . . . He is not in any way guilty of any kind of homicide. It was all my doing . . . Please do not let me die with the guilt of Eddie Hill's murder convictions."

The Parole Board was unmoved. "If Mr. Hill is innocent," asked a prosecutor, "what kind of man would allow him to sit in prison for 23 years of a life sentence?" Eddie Hill's petition for reduction of his sentence was declined.

WITHIN THE FLAT FENCED ACRES and locked-down buildings of the Warren Correctional Institute, in which time itself seems imprisoned, the 4 Paws class, twice a month, is like the circus coming to town. Jeremy Dulebohn typically arrives with an entourage of assistants, new dogs on leashes and in crates, and a returning dog in need of brush-up work from the remarkable Eddie Hill. Jeremy's own enormous black-and-tan German shepherd dog, Brody, canters beside him. Together, even in the corridors of a prison, the soft-spoken man and the rippling dog radiate power.

In khaki slacks, black sneakers, and a gray polo shirt with the 4 Paws logo, Jeremy assumes a wide-legged stance in front of the class and rubs his hands together to get started. The inmates find their seats and call over their pups. They correctly surmise that their interactions with the dogs are closely monitored from the moment Jeremy enters the room, and they're anxious to do well and to stay in this program. The dogs, on the other hand, have no clue anything's at stake.

"Down," the men say, and some of the dogs come over and drop down, but others skitter off sideways with wide-open mouths to snare a few more minutes of playtime. Jeremy waits until everyone has calmed down and then spends a long moment looking into every face in the room—unique in the animal world, dogs seek eye contact with humans—giving every pair of eyes (brown, blue, amber) a moment to rise and flicker across his own. Jeremy appreciates the inmates' honest work and their willingness to listen and learn. But it's not their moods or states of mind that interest him; he wants to see how the dogs are doing. Most love it here. "Who *wouldn't* want to be a dog in prison?" Karen says. The nonstop company of their humans (they sleep in their handlers' cells at night) and of other dogs is heaven to them.

But occasionally a pup comes along who *doesn't* want to be a dog in prison. A pooch with no grasp of the basics ("Sit," "Stay," "Heel," "Come," "Down") after two weeks?—well, it happens. Still inconsistent after four weeks?—a bit of a concern for Jeremy. After six weeks?—red flag: this dog is not learning with this handler. More troublesome: a dog flinching at the sound of its handler's voice, or displaying timidity or stress—ears flattened, head lowered, tail between legs, avoiding eye contact, or slinking to hide under a chair.

The volunteers here are screened by prison administrators, but mistakes can happen—someone with a short temper might slip through; or the chemistry between a man and dog may be off. Someone may have applied to 4 Paws for its novelty factor, for a break in the imposed boredom, without a real desire to connect with an animal. A papillon or German shepherd, famous for loyalty, may mourn the loss of its favorite trainer back in Xenia. Or the prison environment may crush a too-sensitive soul. Jeremy is the occasional recipient of desperate clues during class time, a mute appeal rising from worried eyes, crumpled eyebrows, or trembling legs. Whatever the reason, Jeremy will take that pup with him when he goes.

From small-town Ohio like many of these men, Jeremy offers them a square deal: they fork over hundreds of hours to 4 Paws from their warehouse of infinite time; and he teaches them the craft of winning the trust, respect, obedience, and affection of a dog. The chance to acquire a skilled trade is embraced by some as a practical step toward a future job in the outside world. (One parolee was recently hired to the dog-care crew at 4 Paws.) For many, the rewards go deeper, as the relationships with their dogs are the most affectionate, gentle, and reciprocal they've ever known.

Karen recently received this emailed message from a stranger, regarding a 4 Paws golden Labrador retriever: "Hello, Karen. I hope you are doing well. My fiancé, Joshua, is currently training one of your puppies at Warren Correctional Institution. He and his cellmate have Chantilly and they are SO in love with her . . . I want to thank you and your organization from the bottom of my heart. This is the happiest I have heard Josh in years. You are making such a big difference in his life. Chantilly is now truly all that he talks about."

"Let's see what you've got for me," says Jeremy, and—as the first man-and-dog pair enter the circle to demonstrate walking-on-leash, sitting, and staying—the session begins.

EDDIE HILL, FORTY-NINE, HAS SERVED twenty-five years in prison and has thirty-five years more to serve. He has auburn hair, a chunky nose, an auburn goatee, and a shy, self-deprecating manner. To Karen Shirk, he looks like an accountant or bank teller. He deflects praise by squeezing his eyes shut and throwing his head back at an angle in demurring laughter. And there's plenty of praise to contend with: Eddie Hill is *the* go-to trainer for 4 Paws, possessing the perfect combination of instinct, empathy, animal intelligence, and kindness. "Skills it takes the other handlers four to six to eight weeks to show me with their dogs, Eddie Hill's got them doing at the two-week mark," Jeremy told me.

Karen recognized the gift almost immediately. As soon as Jeremy joined 4 Paws and took over the prison classes, he saw it, too. They could ask Eddie to demonstrate this or that subtle bit of handiwork with a dog and it was executed perfectly; they could place difficult or puzzling or recalcitrant dogs with him and he unlocked their secrets. They began asking for more sophisticated training from Eddie's dogs, not just basic obedience, but advanced service dog skills, like turning on lights, opening drawers, opening doors. Prison guards pulled Eddie aside for private consultations about their own dogs at home. Though it was tricky for a prisoner to admit any kind of weakness to another prisoner, other inmates came to Eddie for help with their 4 Paws dogs. And if any non-Eddie-Hill-trained dog proved problematic after its placement with a family, Jeremy brought the animal to Eddie in prison for brush-up work.

It was more than obedience Eddie won from the dogs—it was their freely given tail-wagging love, eagerness to work, and happiness. After bunking with Eddie for a few weeks, a dog bounded into class, brimming with high spirits, eager to show off. Eddie Hill was not only the most gifted natural-born dog trainer that Jeremy Dulebohn and Karen Shirk had ever seen behind bars; he was among the best they'd ever known.

EDDIE HAS TRAINED SEVENTY DOGS since Timber and keeps a detailed journal of every one. "Timber and every dog since then have given me back everything I've put into them tenfold," he told me when we met in a small office near the prison classroom. "Since Timber, I just look in their eyes and I see them in there. Sometimes they're way back in there, scared to come out; or sometimes they're way to the right or the left of where they ought to be—too scared and submissive, or too dominant, or too insecure. I want to help them get centered. I have loved meeting every one of them. Every one of them has taught me something new and important. They're all individuals.

"There's no chance in here for the closeness with people like the closeness and total giving that is possible with the dogs. People here are very hesitant to even touch unless it's like a handshake or fist-bump. When somebody is leaving forever, you might give each other a hug and that's a big deal. With the dogs, there's hugs and kisses galore and total trust, almost right away.

"I wish I'd known about this before I ended up in here. If I ever get out, it's sure what I want to do, but I'm not up for parole until 2049."

"We'd hire him tomorrow," says Karen. "We keep writing to the Parole Board and to the governor. He's something special."

But few applicants to the Parole Board receive hearings in the Ohio correctional system. Only one case in more than 550 has warranted a hearing in two and a half years. Jessie Balmert, of Gannett newspapers, notes: "That has prisoners and their families questioning how thorough the reviews are."

Most recently, the Ohio Parole Board voted 6–0 not to recommend a commutation of Eddie Hill's sentence to the governor, saying: "The applicant is currently incarcerated for very serious offenses, which include the shooting deaths of 2 male victims. Nothing presented in the clemency application as mitigation outweighs the seriousness of the offense."

At this rate, Eddie Hill's sentence will not be reviewed until 2049, at which point he will be eighty-three years old.

WHEN JEREMY FIRST SHOWED UP with a few papillons in tow, the men in the WCI class scoffed. "I told my cell-mate, 'He better not give *me* one of those frou-frou dogs!'" Eddie Hill told me. "Sure enough, that's what he gave me. I was like, 'This ain't no dog. What am I supposed to do with this thing?' And the other guys are looking at you, like, you know. But I found out pretty quick: that was one smart dog. She impressed the hell out of me. There wasn't nothing she couldn't learn. After that, I told Jeremy, 'From here on out, you can bring me as many papillons as you like.'"

Eddie trained Dante, a papillon, to assist a boy with pervasive developmental delays; and Keeper, a black Lab who attends to a child with bipolar disorder; and Kita, a German shepherd dog, for seizure detection; and Embry, a German retriever, who became a mobility dog for a child in a wheelchair; and Jiminy, a black Lab, who lives with a family in which the mother has amyotrophic lateral sclerosis (ALS) and the child is on the autism spectrum. He trained Micah, a papillon, to become a hearing-ear dog for a teenage boy who'd said he didn't want a service dog because he only liked cats but then accepted a papillon as close enough to a cat and now adores Micah and goes nowhere without him. Eddie also trains PTSD dogs for veterans.

More than a hundred men incarcerated at WCI have followed Eddie Hill's lead. Their magnanimous work with the dogs behind bars has tremendous resonance in the outside world. Fragile children and returning vets will be rescued again and again and again by these dogs. Lives will be saved. Parents of children with 4 Paws dogs write to the prisoners to thank them, each note a gift of incalculable value.

As today's class concludes, the dogs are released for a last round of snorts and wiggles and sniffs. A transfer of dogs is made—a few men greet their new charges and hear details of their assignment—it's an exciting time for them; a few other men kneel to stroke, hug, and kiss their dogs goodbye forever. Through tears, heads lowered, these men accept the news from Jeremy about which dog he'll bring them next time.

The tallest, oldest prisoner—the bearded black man with the iron biceps—bends to lift his silky papillon onto the high shelf of his folded arms. She's not leaving yet, and he is content. As he exits the class in long powerful strides, his falsetto voice rings out: "Say 'bye-bye!' Say 'see you next time!' "

CHAPTER 17

Micah & Casey

D eb Millard tried to make an outing out of taking Casey
for a walk—"Shall we go to the playground, girl?!"—but
it felt to her like Casey was just being polite. She took
Casey in the car to the market. In the parking lot she turned to let

Casey out the back door, then stopped . . . Casey wasn't wearing her red service dog vest. You couldn't walk into a grocery with just some random dog. Casey wasn't a service dog anymore! Deb collapsed in the backseat, hugging Casey, and cried.

Casey dropped in weight. She looked gaunt. She nosed her kibble dispiritedly and only occasionally chewed. One morning Deb heard a gagging, heaving sound; she looked into Connor's room and found Casey retching all over the floor. "Casey, come on now! Outside!" She led the dog to the back door and returned to clean up the mess. The next day Casey threw up again on Connor's floor, all over the little football rug, and then it became a daily occurrence: she choked down a few nuggets of kibble and then upchucked on Connor's floor. Deb had to throw out the precious rug and keep Connor's door closed to Casey.

Half a year passed. Deb and Scott launched, in Connor's memory, a support organization for families of medically complex children called Connor's House. Online, on the phone, and in person, Deb helped parents navigate the maze of hospital systems and medical terminology and health insurance claims and, sometimes, end-of-life decision-making. She hosted blood drives and meetings and fund-raising projects for families without insurance or whose insurance ran out.

Casey stopped pacing. She stayed on her bed. Head resting on paws, she kept her muzzle pointing toward the front door, but without hope. The light had gone out of her eyes. She was a thin, dull, quiet dog now. The two people in the house offered tender solace to each other, but, despite their best efforts to include her and talk to her and brush her and walk her, they couldn't seem to relay sufficient solace of a canine kind.

ON NOVEMBER 6, 2010, HEARING the car pull into the driveway, Casey did a rare thing: she got out of her bed to greet the Millards. She was six years old now, but stood stiffly, like an older dog. The wild copper sheen of her fur had flattened to a dusty mushroom-colored coat, like an old rug. She didn't dance about in the joyful wagging silly ways of her short happy life with Connor. She didn't streak between the window and the front door as if it were one of the historic and glorious days that Connor came home from the hospital. The moment the car turned in from the street—perhaps the moment the car rounded the block—Casey knew Connor wasn't in it.

But on this day—three years and forty-nine days since Connor, lying on the den sofa between his parents, somehow went missing—something was different, worth getting out of bed for. Deb and Scott entered the front door in sky-high moods, their voices marbled with laughter. Deb held a blanket-wrapped something—*someone?*—in her arms, while Scott banged across the threshold with equipment (a molded plastic seat, a plastic bathtub, and a diaper bag rattling with squeaky-clean plastic bottles).

Small people who required equipment were Casey's *favorite*.

She approached cautiously, noncelebratory but curious. The Millards' unusual buoyancy piqued her interest, as did the intuition that, after 1,114 days of the three of them alone in this silent tidy house, there was suddenly, inexplicably, a fourth.

"Look, Casey, come here, girl," they said, kneeling down—Scott, thirty-six, whose face and body had gone lean and hard with grief, was a marathon runner now; Deb, thirty-nine, had grown somehow softer, paler, and gentler over time, with the kindest of melancholy smiles. There was no one on earth they were happier to introduce to their new son than this disheveled lonely dog.

"Casey," said Deb, "meet Micah. Micah, this is"—and here a sound caught in her throat for a moment; had she been about to say *your* dog?—"Micah, this is Casey." The swaddled baby, with a high forehead, soft gray eyelids, and shapely raspberry-colored lips, slept. Casey sniffed the air around the newcomer, a pungent complex bouquet. She sensed the people's happiness and excitement; she felt a sea change. But Micah wasn't Connor, and Casey was Connor's dog. She went back to bed.

THE MILLARDS HAD WADED THROUGH the deep marshland of grief for two years, and more. Occasionally, well-meaning friends and fellow congregants asked, "Are you guys thinking about adopting?" but the Millards could hardly decipher the meaning of such intrusive and bizarre questions. They had no thought for, or heart for, any child but Connor. They were Connor's parents.

But, in time, it seemed possible that Connor—in heaven, or posthumously—wouldn't object to a small sibling.

They learned that it was healthy for an adopted child to retain contact with his or her birth family, so they applied to a local agency offering open domestic infant adoptions. Scanning the list of special needs, the Millards saw nothing they couldn't handle. They checked yes on genetic syndromes, physical anomalies, and prenatal drug exposures. There was only one issue, really, they hoped to avoid. They needed to be excused, this time, if it wasn't too great an inconvenience, from becoming the parents of a child with a terminal illness.

They gave an enthusiastic yes when offered a baby destined to be born with dwarfism, and they excitedly studied and prepared. After his birth, his parents reconsidered, and kept the child, and the Millards slid back toward bereavement. Then a single African-American woman, and her mother, in search of an observant Christian home, chose the Millards. "Keep in mind that you'll attract attention; strangers will stare at you," the adoption agency staff member cautioned them, prospective white parents of a black or biracial baby.

The Millards nodded dutifully. They'd never experienced a single day of parenthood, in Connor's eight and a half years, when they hadn't been stared at, judged, pitied, commented on, studied, examined, and boldly questioned.

"Stared at?" they replied. "Oh, we can handle staring."

On November 4, 2010, Micah was born full-term, flawless, gorgeous. Given a clean bill of health, he was dismissed on his second day, and came home in the arms of his new parents. Micah was *fine*.

The Millards were mostly fine. The Millards were confused. Like Old Faithful, grief for Connor burst out of the ground under their feet. And everything they knew about infancy, about parenthood, was wrong. They hesitated, at night, to place the baby in a plain crib unencumbered by machinery—was he supposed to breathe *on his own*, all night long? When they remembered sudden infant death syndrome, they panicked. SIDS was about the only danger Connor *hadn't* faced, since a machine breathed for him. Anxiously they researched other health issues facing newborns. Crib bumpers were out of favor. Babies should sleep on their backs. Circumcision was debatable. They researched every issue in depth, but the answers seemed . . . strangely simple. Most websites for new parents focused on baby names and shopping tips, rather than issues of life and death.

The Millards turned to their friends for advice. Touched by the naïveté of the questions—"Are we supposed to let our doctor know if the baby gets a cold?" "Is baby abduction something we need to be really careful about?"—the friends smiled kind of sympathetically and said, "Really?"

In love with the baby, awed by the baby, Deb and Scott together tended to Micah in his bath, and on the changing table, and in his crib. They stood above him as he slept. When Scott said, "This is mind-blowing," Deb knew what he meant: *This is so easy it's terrifying.*

One day, when Micah was about a month old and Scott was at work, Deb needed to run to the grocery. She called her mother to ask her to watch Micah for a few minutes as she used to do with Connor.

Deb Tiel said: "Take him."

"*Take* him!?"

"Take him."

Deb took him. She laid Micah gently in his car seat and he cooed. She pulled him from his car seat and slid him into the baby sling on her chest, and he looked around with soft brown eyes. The cashier exclaimed over his beauty. Back in his car seat, he smiled up at Mama and pedaled his fat legs. And yet, driving home, Deb started sobbing and then, at a traffic light, crying hard. She didn't know if she was crying from relief, or love, or amazement, or bliss, or grief. Mostly happiness and love. Also grief.

AS FOR GETTING LOOKED AT, it wasn't much of an issue. Inter-married couples and mixed-race families weren't rarities in New Jersey. But friendly conversational gambits could turn painful. When strangers or acquaintances enthused over the jolly tawny baby—now sprouting a halo of soft brown curls—they sometimes asked, "Is he your first?"

How well do I know this person? Deb would think. "Not my first," she'd reply gingerly. "I have an older son." Then she would bustle away as quickly as possible.

When out with a group of new friends, getting acquainted, the Millards grew very still when the chitchat turned from sports, church, or work to family. They tried to change the subject, avoid their turn. Inevitably someone asked if they had kids.

Scott thought: *A deceased child is a conversation killer, but I'll feel guilty if I don't mention Connor. If I do mention him, it will take us to a deep level really quickly. Am I ready to go there right now?*

Are they *ready to go there?* Deb thought. *Can we just answer on a need-to-know basis?*

If putting their baby to bed in a crib without mechanical extensions reaching into his throat, lungs, and stomach unsettled the Millards, living with a mobile toddler alarmed them more. *He rolled over in his sleep—is that safe?* (Connor had always been pinned in place by his equipment.) *He's not connected to anything!* they realized when Micah started crawling. *He can go anywhere he wants! He could, like, just disappear.* And disappear Micah did, like the world's healthiest toddler. One second he was on the floor of the kitchen, at Deb's feet, opening and closing cabinets and banging on pots, and the next minute . . . he was gone! He could make it all the way down the hall in a flash! They'd find him standing up against the back door, laughing at squirrels. For fun, the Millards picked up books on American Sign Language for babies and taught Micah signing. At twelve months old, he could show them the signs for "sleep," "milk," "more," and "all done." He learned to shape the letter "C" for "Connor," the name of the boy in framed photographs around the house. One day, as Deb and Scott reminisced about Connor, they glanced over at Micah in his circle of toys at their feet and discovered the little fellow, who'd been listening to them, holding up "C."

MICAH LAUGHED WHENEVER THE BIG dog galumphed into the room. But the dog didn't seem to return the laughter. Casey had exhausted most of her interest in Micah after the first few days. Even the thickly aromatic diapers were repetitive after a while, not worth the trip down the hall to the nursery. Casey surely noticed that the house had filled up with renewed activity and laughter; that Deb was cooking again; that people came over to visit and Scott grilled hamburgers on the deck. Everyone who entered the house greeted Casey. Some stroked her head, looked into her dark eyes, and spoke soothing words: "Poor girl, good girl." Casey got out of bed willingly now to accompany the Millards on walks with the stroller. She didn't mind the baby; she just wasn't interested in him.

Around Micah's six-month birthday, though, something about the baby changed, attracting Casey's attention. Micah started eating solid food with fascinating aromas, textures, and flavors. Also: Micah was messy! The floor around Micah's high chair filled up like a primeval landscape under a meteor shower. Scrambled egg dropped from the heavens! Bits of cheese! Cubes of soft meat! Boiled carrots and green peas! Noodles! Cheerios! Globs of applesauce! And gnawed-on damp Zwieback teething biscuits! *I don't know what it is about this kid, but when he's around, good things happen!*

Casey became a regular attendee at Micah's feeding events. As Deb slid Micah into the high chair, Casey's tail ticked like a metronome. When Micah peered over the edge of his tray and found the shaggy dog looking down, looking up, looking down, the baby giggled and threw her a gob of spinach soufflé. Casey gobbled up the prize, licked her lips, and stood at attention, hoping for another. The baby threw a missile of cheese. The dog bounded to get it, gulped it down, and returned. The baby pounded on the tray, flinging food bits like confetti. In this way, a friendship was born.

When he was removed from the high chair, Micah was covered with fascinating smears and dribbles. Casey could help with that! Micah let the dog lick him all over, even though the hearty licks knocked him from a sitting position into a lying-down position. From the floor, Micah had a better angle from which to examine something that had long intrigued him: Casey's jingling collar tags. Casey stood still while the fat greasy little fingers fumbled at the tags. Deb intervened when Micah began to pull the tags toward his mouth in order to gnaw on them.

Deb took Micah and Casey onto the deck, and into the backyard. There, one afternoon, a remarkable thing happened: Casey barked at a squirrel. Since the night Connor's baby monitor had run out of batteries, Deb hadn't heard Casey bark. She'd half-thought Casey *couldn't* bark for some reason. But, starting that day, Casey barked as often as any other dog—she barked at birds, at the UPS man, at other dogs. Casey could bark. She had *refrained* from barking for the length of Connor's lifetime because barking startled Connor. Now she was becoming a regular dog.

She began showing up at Micah's bedside when he woke up in the morning and wagged when Micah opened his eyes. In the living room, Micah discovered that, if he threw a toy, Casey would fetch it. One day Micah sniffed at Deb, and that sniff—sounding just like Casey's sniff—she understood to be Micah's word for "Casey." Sometimes Micah stood at the back door looking out, watching Casey roam around the yard. When she barked, he laughed.

One day Micah couldn't spot Casey in the backyard. He turned to Deb, sniffed once, and covered his eyes as in the game of peek-a-boo. She understood him: *Casey . . . Hiding.* He was a happy boy. Though she and Scott would never stop longing for Connor, treasuring every moment of life with Connor, their lives had regained sweetness.

Life is pretty sweet for Casey now, too. She likes Micah a lot; she may even love him. Wherever Micah goes in the house,

Casey follows. When he gives the command for "Sit" or "Down," she promptly obeys. Micah giggles every time.

"How is Casey doing?" friends occasionally ask the Millards.

"She's better, she's good," Deb tells them.

"She's mostly just a dog now," Deb says to me. "She's almost eleven. Micah is afraid of animals in general, but he's not scared of Casey. He'll sit with her and hug her, although it scares him when she runs by him really fast. He is little and she is big in his mind! She'll lie by his bedroom door when he's sick. But . . . it's different. I think she's getting really attached to Micah, she's having fun with Micah. But Casey was absolutely and most definitely *Connor*'s dog. And when Connor died, her life changed so much, she has never really been the same. The sadness of missing Connor is still in her eyes, and in her walk. I think she feels it deep inside her. I think when you love someone that much, the love stays with you and is always part of you. I can't imagine anyone wanting it any other way."

CHAPTER 18

Iyal & Chancer

I s your son likely to verbally abuse a dog?" Karen asked a
mother during an intake phone call in 2007.

"Oh! Well, oh! Verbally abuse? Actually . . . I guess
I would say, I guess I would have to say . . . yes," said Donnie
Winokur, phoning from Atlanta. Until that question, she'd been

quite the fast talker in explaining her family's situation, gaining momentum as she sensed in Karen a kind listener, gaining volume as she intuited that, after years of pandemonium at home, she might have stumbled across a lifeline. On and on she talked in words perfectly clear and well enunciated, effectively streaming along in 12-point Times New Roman font.

Donnie was a skilled public speaker now, having become a statewide advocate for her son and others like him. Her smart cascade of words prevented—or, at least, delayed—the fast-approaching moment when this service dog agency director would tell her NO.

Because, in truth (it occurred to her, while trying to cajole her way into the director's heart), if this idea of a service dog for Iyal didn't pan out, an idea she'd been mulling over for months with mounting conviction, she was out of ideas. So she kept talking, promoting, soft-soaping, anything to avoid having to hang up the phone and turn to face her lovely house, which had at its heart a kind of howling loneliness. The sparkling-clean kitchen and tall foyer and modern art on the walls, the perfect suburban cul-de-sac home into which they could never invite anybody . . . Then Karen Shirk's question stopped her. And Donnie told her the truth, that her son—who screamed at all of them day and night—would probably scream at a dog, too.

"Is your child physically aggressive?" asked Karen.

There was a long pause, after which Donnie sighed and whispered: "Yes."

"When he's aggressive, is it randomly directed at people with no rhyme or reason, or is it more avoidance/defensive? Like you want him to eat his dinner with a fork and he doesn't want to? Or you want him to wear clothes and he wants to be naked? Is the aggression more aimed at someone who's asking something of him?"

"Yes, more like that. Avoidance," said Donnie. "Or just acting out, not directed at anyone. Like the world's just putting too much pressure on him."

"Got it," said Karen. "That doesn't concern me. That type of aggression is virtually never aimed at an animal. We'll ask you to intervene and inform us immediately if you ever see aggression toward a dog—obviously we don't want our animals to be harmed—but I do not think you will see any."

CHANCER, A ROLY-POLY PUREBRED GOLDEN retriever puppy, was born at Mervar Kennel in Youngstown, Ohio. When he was old enough to leave home, he was purchased by a family of four: two parents, two school-age kids. It looked like the makings of a happy dog life, Judy Mervar felt, watching the family drive away. She always hoped for the best. For the first few months in his new family, Chancer was kissed, doted upon, and carried about like a plush toy. He was probably happy, but he got no discipline, no structure, no regular exercise, no obedience training. When he grew too big for the children to carry, their attention wandered to other pursuits, and suddenly Chancer, on his own in the house, began making mistakes. He got yelled at, jerked by the collar, swatted with a rolled newspaper across his tender nose, and ultimately deposited in a small fenced-in enclosure with a doghouse in the backyard. There he lay for long hours with his face on his paws. He wagged sadly when a member of the family came out to pick up his empty bowls and bring them back with kibble and fresh water, stroking his head and saying kind words. But he no longer wiggled all over in anticipation of being taken back to the house. He grew fat. Surveying him through the kitchen window one day, the adults in the family decided to give him up. Their contract gave Judy Mervar first right of refusal.

"I've seen this before," she told me. "People buy an adorable puppy and don't train it and then they don't like it anymore. When Chancer was a year old, these people called and said they didn't have time for the dog, the kids weren't helping, could they bring him back." She was shocked when she saw Chancer. "Overweight. Filthy. No training, no manners. He couldn't even walk on a leash. They had done nothing for this dog but put a bowl of food in front of him. He was still a lovable sweetheart with a great temperament, but he was an undisciplined mess." She phoned Karen Shirk. "Karen loves our dogs. She knew Chancer's parents—she's bought puppies from the same line. We're three hours away. She sent someone right away to pick him up."

In six months at 4 Paws, Chancer slimmed down, cheered up enormously, made lots of dog friends, and took to his training with zest. He was closer now to the gorgeous, bright-eyed, engaged, prancing animal he was born to be. He was physically strong. And, despite everything, he had naturally high (Jeremy felt) self-esteem. A trace of doghouse-loneliness lingered in his soul. But that was by design. It wasn't yet time for Chancer to fall in love again.

Dogs at 4 Paws are shown great kindness but, once the dogs have outgrown puppyhood, the staff refrains from giving them wholehearted, best friends intimacy. "We pet them and love them, but we don't give them that intense, *I love you so much, you're my baby,* kind of one-on-one attention," Karen told me. "After puppyhood, we don't take them home with us at night. Every one of our dogs wants that closeness, is primed for that closeness, but we need them to save it for their families."

Chancer probably experienced a kind of blankness in those months of 4 Paws training. He couldn't know that anything was missing, that this wasn't his real life, that he was on deck. At two years of age, he was still a dog-in-waiting, without a best human friend to love him more than anything in the world.

"Chancer," Karen said, "desperately needed a boy."

"OKAY," KAREN SAID TO DONNIE ON THE PHONE. "Tell me more."

"You're going to give us a dog," Donnie replied, too surprised to curl the words up into a question.

"Well, I've never heard of FAS, but your son's issues don't sound all that different from a lot of our kids. I don't see a problem here."

A founder of the Georgia affiliate of the National Organization on Fetal Alcohol Syndrome (NOFAS-GA), Donnie now forged ahead with rapid and precise diction. Iyal was nine years old, his IQ was eighty and falling, his language was rudimentary. He got hooked on bizarre thoughts and repeated them endlessly. He was incapable of making decisions. He still suffered from night terrors and bedwetting. He seemed to care little for his family, or for anyone else. He seemed alone in the universe. Iyal's doctors had tried twenty different medications, including atypical anti-psychotics, without lasting success.

"Let me tell you how to get started," said Karen. "There's paperwork, there's fund-raising, we need video. We have people who can help you through it."

"You're going to give us a dog," Donnie stated again in flat disbelief.

"If your son and your family can be helped by a dog and you can take good care of a dog, we're going to give you a dog," said Karen Shirk. With this credo she had launched 4 Paws; with this credo she threw open the doors to families in need. Today, nearing placement of her thousandth dog, to this credo—and its implicit compassion and respect for suffering people—she has remained faithful.

When Karen said yes to Iyal Winokur, it signaled another breakthrough for 4 Paws. If this placement worked out, the dog was likely to be the first Fetal Alcohol Spectrum Disorder [FASD]-assistance dog in the world.

"I'm amazed by everything," Karen told me recently. "How big we are, how many different things we do, the fantastic client-base we have (about half stay in touch pretty frequently through social media), the respect in the field we have earned. Agencies that scoffed at us for placing dogs with children now place dogs with children."

"WAIT. YOU WANT TO BUY a thirteen-thousand-dollar *dog*?" asked Harvey that night, almost unable to believe this topic hadn't been put to bed. "You want to spend that kind of money, which we don't have, on a *dog*? Instead of for a nanny, or respite care, or private school? What sense does that make? And where are we supposed to get thirteen thousand dollars?"

"My father will help us," said Donnie quietly. She felt at peace with this decision; she felt steely and cool.

"A dog won't mean anything to Iyal," said Harvey.

"It might. And imagine if it did. There would be one living thing in the world for Iyal to care about."

"You're talking about a dog with a vest like a seeing-eye dog? It will be embarrassing to even go into public like that."

"It's already embarrassing to go into public with Iyal."

"A dog in a vest will make him seem so disabled."

"A dog in a vest will tell people that he acts like this because he's living with a disability," she said with precision.

"Donnie," he moaned. But he could see that his wife had

completed her research and was making her stand. She wanted this dog. With her crisp manner around the house, Harvey got the message that he'd better not overdo the "It's me or the dog" threats. Still, he held on for two months before conceding. He drew the line at attending the required ten-day class in Xenia, citing congregational obligations.

Among the dogs-in-waiting, Jeremy chose Chancer as strong enough and self-confident enough to handle anything Iyal tried to dish out. In January 2008, Donnie, her doting elderly father, her first cousin, and her two children drove to Ohio.

Donnie had fantasized that the change of scene might inhibit Iyal's explosive behavior—he was throwing one or two major tantrums daily at home. She got the family settled in along a stretch of chairs in the training circle and Iyal ran over to the video game in the play area. Midmorning, Donnie tried to coax Iyal to come watch Jeremy's demonstration of the proper way to hold a leash. It struck her as something Iyal could and should learn. "Look, Iyal! Come here, sweetie!" she enthused in an especially musical, high-pitched, and breathless voice. "We're going to meet Chancer after lunch. Do you think you'll be able to hold Chancer's leash like that? I bet you will be the best one at holding his leash!" As she guided the big nine-year-old by the arm back to the training circle, he crumpled to the floor in the middle of the room and began kicking and boo-hooing like a jumbo toddler. "I'm so sorry!" cried Donnie as the class came to a halt. She knelt beside her son and quietly spoke his name, but he was unreachable, hysterical, kicking his feet and shrieking, "NOOOOOOO!"

"Don't worry," said Jeremy; "take a break, everyone." Special needs parents all, no one judged or disapproved of the scene in the middle of the circle; they simply turned to their own children's needs.

Donnie, looking around the circle with a down-tilted head, through forward-falling hair, perceived that no one here was judging Iyal: he wasn't necessarily the worst-behaved or most challenging child here. Unfortunately, Iyal did seem to be the worst-behaved child in downtown Xenia at lunch that day. When the Winokurs walked the few blocks to town, Iyal lost control again. He sat down hard on the pavement, crossed his arms, and bawled. This happened on the drive-through lane at Wendy's. He'd grown too heavy for Donnie to physically lift and relocate. The backed-up drivers looked at Donnie through their windows far less empathetically than had the 4 Paws parents.

Back in Karen's office, Jeremy said: "Iyal Winokur desperately needs a dog."

A FANTASTIC AIR OF HAPPINESS filled the social hall after lunch. The families were about to meet their dogs! One by one, as people gasped and cameras flashed, the dogs—from golden retrievers to sassy little papillons—sauntered into the room. They seemed to sense they were beautiful, important, and desired. Chancer trotted fifth in line, unaware this was his red-letter day. He glowed with shaggy tawny warmth. As his trainer handed over the leash to Donnie, the dog panted with a wide-open smile. Morasha dropped to her knees and embraced Chancer's big neck. Donnie, who'd never had a dog before, felt like doing the same. "Hi, hi, hi, good boy," she murmured, suddenly choked up, stroking his broad warm head. "Iyal, this is your dog. Say hello to Chancer!" Iyal said, "Hello," without touching him, then walked back to the play area, where he played by himself at a computer terminal.

AT THE END OF THE second day, the families were given permission to take their dogs with them overnight. At a nearby motel,

Donnie's cousin took Chancer outside for a walk while Donnie watched Iyal and Morasha splash in a hot tub in the solarium beside the lobby. "When they came back from their walk," Donnie told me, "Chancer looked around, and then broke away from my cousin. I thought: *Oh my God, he's escaping. We're going to lose him. There goes the thirteen-thousand-dollar wonder!* He streaked past everybody in the lobby, tore into the solarium, and took a flying leap into the hot tub. He was saving Iyal!"

Chancer had not been trained in water rescue.

"Was he saving Iyal from drowning?" Karen laughs. "I don't know. I'm not sure Iyal was in danger in a hot tub with his mother watching."

Over thousands of years, golden retrievers have been bred to leap fearlessly into bodies of water to collect waterfowl. Maybe the bubbly hotel hot tub looked to Chancer like a very small lake in which he might locate a duck.

"Chancer probably just wanted to play," Karen says. "Thirty-six hours after meeting Iyal, Chancer would have been bonding like crazy and ready to have some fun."

The reverse may not have been true. The havoc wreaked by alcohol on a developing child's brain includes scrambling the emotional pathways. The routes to friendship, fun, closeness, and love may be underdeveloped or buried under cognitive roadblocks. Still, Iyal's burst of laughter when the big yellow dog came sailing through the air and clumsily exploded into the hot tub was the greatest sound his mother had heard out of him in a long time.

On their first night together, the people went to bed and Chancer laid down on his mat on the floor as instructed. But sometime in the night he must have jumped up to join Iyal; and sometime in the night, Iyal must have agreed to share his bed, because that's how everyone found them the next morning.

"THE MOMENT THEY CAME HOME to Atlanta and walked in the house with Chancer, I knew something had changed," Rabbi Winokur told me. "I could feel it instantly, the magnetism between Iyal and the dog. Children affected by brain damage or psychiatric disorders have trouble feeling close to anyone, and they rarely feel safe. Iyal had been pretty lost in the world. You could see right away, from day one, that Chancer grounded Iyal. Chancer was going to be his emotional and physical anchor."

The morning after Chancer's first night in Atlanta, the Winokurs woke up in their bedroom from a full night's sleep for almost the first time since 1999. They looked at each other in semi-horror: Was Iyal still alive? There had been no 2:30 a.m., 4:00 a.m., 5:15 a.m. disruptions. They found Iyal snoozing in his bed beside the big yellow dog, the latter hogging the mattress. And that's how their nights would remain. Iyal might wake up with night terrors, but Chancer evidently calmed him and lulled him back to sleep. The Winokurs joined a large chorus of 4 Paws parents who say: "If the dog had been nothing other than a sleep assistance dog, he would have been worth every penny."

"IN A VERY SHORT TIME, Chancer became incredibly attuned to Iyal," Donnie said. Today, the dog and boy are like E.T. and Elliott: when Iyal is happy, Chancer is happy. When Iyal is distressed, Chancer is distressed. Unlike Iyal, when Iyal is distressed, Chancer knows what to do about it. When Iyal's opening roar is heard in the distance, Chancer bounds through the house to find him. *Game on!*

Iyal typically rages by crossing his arms, sitting down hard on the floor, and screaming and kicking. Chancer unknots the crossed arms by inserting his wide friendly muzzle through the locked arms from below, opening them up, and nuzzling toward

Iyal's face, licking and slobbering, until the boy's screams turn to laughter.

In the early weeks, Chancer disrupted Iyal's meltdowns when commanded to do so: Harvey or Donnie said, "Lap" or "Nudge." By the second month, he interrupted meltdowns on his own, without needing a command; and then he began heading off meltdowns before they began.

Tutors and occupational therapists make house calls for Iyal; they do their work in the living room while Chancer dozes on the polished wood of the foyer. He twitches in sleep, his paws paddling through ancestral streams in pursuit of ancestral waterfowl. But he will suddenly wake up, raise his head and listen . . . or sniff . . . and then he will stride into the living room and insert himself between the tutor and Iyal, or between Iyal and the coffee table upon which the teaching materials are displayed. Chancer stands there stolidly, not distressed, not particularly interested, but unmovable. The tutors and therapists understand the message: *We're done for today. We'll see you Thursday.* They know they ignore Chancer's warnings at their peril.

Iyal may be upstairs in his bedroom and Chancer sprawled on the kitchen floor while Donnie makes dinner, when he suddenly alerts, flicking his ears, tuning in. Sensing that Iyal is about to go haywire, he gallops up the stairs to find him, playfully head-butts and pushes him down to the floor, gets on top of him, stretches out, and relaxes with a satisfied groan. Helplessly pinned under Chancer, Iyal resists, squawks, screams for his mother, and then relaxes, too. Maybe it's a kind of deep muscle massage. Maybe it's just a big dog lying on top of the boy he loves, and sealing off the boy from the dizzying and incomprehensible world for a while.

"We trained Chancer to disrupt Iyal's tantrums," Jeremy says.

314 – The Underdogs

"Being able to *prevent* tantrums is coming from subtle training within the family. He may be reading Donnie's body language or facial expressions, or he may be smelling some chemical changes in Iyal or hearing some noises from him that predict a tantrum. He feels rewarded when Iyal stabilizes."

"The dogs are trained to disrupt a behavior when the parents say, 'Lap' or 'Touch' or 'Nudge' or 'Kisses,' " Karen says. "Then, some of the dogs start to kind of signal Mom or Dad, as if they're asking: *Hey, do you see what's happening here? Do you see what's about to happen here? Give me my cue to go to work.* Over time, about eighty percent start disrupting an inappropriate behavior on their own, without needing the command."

"Lately," Donnie says, "and this is the best yet: if Iyal gets distressed, he goes to find Chancer and he curls up next to him. He picks up Chancer's big paw and gets under it." Many children learn to self-soothe in infancy. Iyal did not. Turning to Chancer to stabilize him is a huge step toward mood self-regulation for this boy.

Two weeks after Chancer's arrival, Iyal startled his parents by using multi-syllabic words. He was also suddenly possessed of opinions, judgments, and important questions, and he expressed them. Donnie says: "B.C., Before Chancer (which is how we refer to our life then), Iyal echoed Morasha word-for-word. It drove her nuts. Every morning I ask, 'Do you want to take your lunch today or eat lunch at school?' and every morning Iyal parrots whatever Morasha says. If she says, 'Take,' he says, 'Take.' If she says, 'Turkey,' he says, 'Turkey.' It drives her nuts. With his frontal lobe damage, decision-making like that is difficult for him. But one morning 'A.C.,' I asked about lunch, Morasha said, 'School,' and Iyal said, 'I'd rather have lunch from home than a school lunch because I think that will taste better.' "

Donnie froze and turned slowly to look at her son. Not only had he rarely expressed an independent opinion, he'd almost never explained his thinking. "It was a level of improved executive function we'd never seen before," Donnie told me, "and a more sophisticated expression of his choice than we'd ever heard.

"B.C., driving in the car with Iyal, if I turned down an unfamiliar route, he might say, 'What happened?'

"A.C., sensing I'd taken a wrong turn, Iyal asked: 'Were you distracted by Chancer and that's why you made a bad turn?' That showed an understanding of cause-and-effect, and a high-level word choice.

"B.C., Iyal never mentioned his disability, although we have educated him about it. Suddenly he started asking us things like, 'Did Chancer's birthmother drink alcohol?' and 'Does Chancer have a boo-boo on his brain?' Then one day he asked me, 'Why did my birthmother drink alcohol?' "

At nine, Before Chancer, Iyal had not yet reached the developmental milestone of "theory of mind"—an insight about other people's consciousness that neurotypical children often achieve by age five. Empathy—the essential ingredient of close relationships—was foreign to him. After Chancer, Iyal began thinking about what *Chancer* liked and what *Chancer* wanted. Only since Chancer's arrival has Iyal shown sheepishness or regret following a tantrum, signaling a new awareness that his outbursts may affect others. "Is Chancer mad at me?" he asks his parents sometimes as he regains his composure, or "Is Chancer disappointed in me?" "Mommy, tell Chancer I love him, okay?" Iyal will ask and Donnie replies, "Honey, you can tell him," so he does.

"The sad flip side of 'theory of mind,' " Donnie said, "is that Iyal is deathly afraid that if he misbehaves too much, Chancer will want to be someone else's dog. We'll be in a park and he'll tell me that Chancer is smiling at another kid and wants to be *his* dog."

THE SCIENCE BEHIND IYAL'S COGNITIVE leaps is still in its infancy. Dr. Alan Beck is among those intrigued by it. "There is a real bond between children and animals," he told me. "The younger the child, the greater the suspension of disbelief about what an animal understands or doesn't understand. The absolutely non-judgmental responses from animals are especially important to children. If your child with FASD starts to misbehave, your face may show disapproval, but the dog's face doesn't show disapproval. The performance anxiety this child may feel all the time is absent when he's with his dog. Suddenly he's relaxed, he's with a peer who doesn't criticize him."

The hypothesis is that the sudden drop in Iyal's anxiety level—the sudden decrease in his hypervigilance, the lowering of his cortisol level, and the disarming of the fight/flight physiology—frees up cognitive energy that he can use for thought and speech. "A child with a disability feels freer not to suppress his ideas and behaviors when he's with his dog," Dr. Beck says. "There's a level of trust and confidentiality he has with no one else. And it's a good choice: the dog *is* his true confidant and friend."

And of course he's intrigued by what the relationships between 4 Paws dogs and their children tell us about the capacity of dogs.

Many of Lassie's rescues of Timmy were improbable, he says, the skyscraper-leaping feats of superhero narratives. But some elements of the Lassie Myth, seen in a new light, appear a bit more realistic today. "Can Lassie fetch the Phillips-head screwdriver?" asks Dr. Beck, referring to his favorite bit of Lassie humor. "No. But if you add the screwdriver to a pile of her toys, would she recognize it as a new item? Yes. If she knew the name of every other item on her pile and you asked her to bring you the screwdriver, a word she didn't know, would she bring it to you? Yes. Some researchers feel that dogs function at about the cognitive level of two- or three-year-old humans. Clearly Lassie was much less cognitively developed than Timmy, but, also clearly, she was smarter than our older views of animals allowed us to see.

"Lassie's intelligence, empathy, memory, object permanence, and face- and voice-recognition skills were not recognized by experts even a decade ago. But that view is changing. As we increasingly appreciate dogs' abilities to recognize the meaning of voices and gazes, and to pick up intentions and motivations from subtle clues, we're seeing that some of Lassie's actions could actually occur in dogs not specifically trained to perform them. It's pretty widely recognized now that animals have thought processes, like people do. They *learn*—it's not all instinct. They have memories. They communicate among themselves. They have lots of senses with which they navigate the environment. There are experiments showing dogs taking advantage of a human looking away and using that moment to steal a treat— that's a kind of 'theory of mind,' an understanding on the part of the dog of where the human's attention is focused.

"Without going too far with this, I think it's safe to say we didn't recognize Lassie's cognitive and emotional capabilities. Maybe it's because our experimental approach was naïve and flawed. Maybe it's because we weren't looking for them."

CHANCER HAS NOT CURED IYAL.

"From the moment Iyal wakes up in the morning, there's tension in the house," said Donnie. "He has neurological and psychological damage Chancer's paws can't reach. But Chancer mitigates the disability. It's like we have a nanny."

Last fall the Winokur family wrestled with the likelihood that Iyal was being bullied at middle school. "Some boys told me to hump a chair," he reported to his mother and psychiatrist a few days into eighth grade.

"Hump a chair?" Donnie said. "I'm not sure what that means."

Iyal stood to reenact it.

"Look at Chancer," murmured the psychiatrist to Donnie. As Iyal engaged provocatively with a chair, Chancer rose, distressed. Whining, he tried to block Iyal's lunges.

On another day, a distraught Iyal told his parents that the boys said: "Go kiss that boy or we'll hang your dog."

"This is the classic setup for people with impaired judgment," Donnie says. "They're at risk of being exploited criminally and sexually. They can become both victim and perpetrator." It may not be one-sided bullying either. "Iyal may have pursued those boys," Donnie says. "He desperately wants friends. He doesn't understand personal space or social distance. He might have been annoying them, and they reacted."

The school principal was instantly responsive; his staff spoke to the other boys. "But Iyal keeps talking about it," Donnie says. "It's hard for Harvey and me to know if the bullying is still happening or if Iyal has just fixated on the trauma of it. Past, present, and future get confused in his mind."

Chancer doesn't accompany Iyal to school because the boy can't take the reins as Chancer's handler. "He can't even take Chancer for a walk around the block," Donnie says. "He might drop the leash, and Chancer might interpret that release as permission to track a hamburger. Chancer's an amazing service dog, but he is a dog, and he loves meat."

If Iyal ordered Chancer to do a wrong or dangerous thing, or to join him in reckless behavior, would Chancer recognize that they were transgressing? Would Chancer disobey Iyal? "When a dog puts a vest on, it changes his persona," Donnie says. "He knows he's working. In the service dog world, they call it the halo effect. Guide dogs for the blind are trained in 'intelligent disobedience' for dangerous situations, like traffic. But I don't know if a dog can reason between right and wrong."

With every passing year, the challenges to Iyal's safety, and to the well-being of those around him, multiply. With puberty, he began trying to grab his mother inappropriately. The Winokurs fear that soon the principal's office will be calling them, rather than vice versa. "Harvey and I feel like we're sitting on a volcano," said Donnie. "Iyal is a thirteen-year-old who functions cognitively, emotionally, and socially like an eight-year-old. That gap will widen. He will never catch up to his chronological age. And most people cannot tell the difference between 'neurological noncompetence' and 'behavioral noncompliance'; in other words, that Iyal's not deliberately misbehaving, that he's doing the best he can."

Iyal will never drive a car. He will never hold a regular job. He doesn't understand money or time. Experts say that the transition from adolescence to adulthood is particularly difficult for individuals with FASD. And Chancer won't be around forever. For as long as they live, the Winokurs hope to make sure there is a 4 Paws dog at Iyal's side; for now, they cannot conceive of a life without Chancer. As he nears retirement, they will bring in a young 4 Paws dog, trained by Jeremy, mentored by Chancer.

Chancer doesn't know that Iyal is cognitively impaired. What he knows is that Iyal is his boy. Chancer loves Iyal in a perfect way, with an unconditional love beyond what even his family can offer him. Chancer never feels disappointed in Iyal or embarrassed by Iyal. Beyond cognitive ability or disability, beyond predictions of a bright future or a dismal one, on a field of grass and hard-packed dirt and crows and dandelions, between the playground and the baseball diamond, you can see them sometimes, the two of them, running, laughing their heads off, sharing a moment of enormous happiness, just a boy and his dog.

Acknowledgments

To the David Black Literary Agency—Jenny Herrera, Susan Raihofer, and the inimitable David Black himself; to Vera Titunik, *New York Times* magazine editor; to Ecco—publisher Dan Halpern, assistant editor Emma Janaskie, copyeditor Lisa Silverman, publicist Ashley Garland, editorial assistant Bridget Read, and my brilliant editor Hilary Redmon; to 4 Paws for Ability—founding director Karen Shirk, head trainer Jeremy Dulebohn, trainers, and staff; to the specialists in many fields who shared their work and insights; to the families who permitted me to tell their stories; and to my own unspeakably brave and marvelous family, I offer respect, gratitude, friendship, and love. I wish I could give each and every one of you a puppy.

Notes

INTRODUCTION

1   **"domesticated dog food":** Sarah Pruitt, "Man's Best (and Oldest) Friend," *History in the Headlines*, May 22, 2013, www.history.com/news/mans-best-and-oldest-friend.

1   **"prehistoric dog-lovers" . . . might have been his toys:** Ibid.

2   **Her breathing changes and her heart rate speeds up:** Phone conversation with Dr. Alan Beck, June 16, 2014.

5   **The Wolfdog seems to have accompanied:** Ken Fischman, "The Tracks in Chauvet Cave," *Ancient Pathways to a Sustainable Future*, January 17, 2012, http://ancientpathwaystoasustainablefuture .org/1huntergatherers/tracking/the-tracks-in-chauvet-cave/.

6   **"The emerging story . . . sees humans and proto-dogs evolving together":** Mark Derr, "From the Cave to the Kennel," *The Wall Street Journal*, October 29, 2011.

9   **"*Octopus* are aneemals":** "Luiz Antonio—A argumentação para não comer polvo," ("Luiz Antonio—Why He Doesn't Want to Eat Octopus"), May 13, 2015, www.youtube.com/watch?v=SrU03da2arE.

9   **"Seventy percent of children confide in their pets":** Phone conversation with Dr. Alan Beck, June 16, 2014.

664442454445646545554666554556654566

CHAPTER 1: JUKE

17   **"catastrophe-prone family":** Denise Gareau, "Angels from God Chronicles: The Lassie Myth," Eskies Online, 2008, www.eskiesonline.com/afgtlm.htm.

17   **an abandoned house, a mine, a wildfire:** Linda M. Young, "Lassie: Frequently Asked Questions," accessed February 10, 2015, www.lassieweb.org/lassfaq.htm.

18   **"Needless to say, no dog is like Lassie":** Alan Beck and Aaron Katcher, *Between Pets and People: The Importance of Animal Companionship* (West Lafayette, IN: Purdue University Press, 1996), 173.

18   **The show "set a ludicrously high bar":** "Lassie: The Perfect Dog Sets High Bar for Real Pups." Ketzel Levine. Morning Edition. Special Series: In Character. NPR. NPR.org. January 07, 2008. http://www.npr.org/templates/story/story.php?storyId=17894690. Web 3 May 2015.

18   **"LASSIE GET HELP":** Shanahan, Danny. *The New Yorker.* May 8, 1989. Cartoon. http://www.condenaststore.com/-sp/Lassie-Get-help-New-Yorker-Cartoon-Prints_i8542025_.htm

19   **Timmy never actually fell into a well:** Young, "Lassie: Frequently Asked Questions."

22   **"complex organic machines":** Garth Kemerling, "Descartes: A New Approach," *Philosophy Pages,* last modified November 12, 2011, www.philosophypages.com/hy/4b.htm.

22   **"eat without pleasure":** Simon P. James, *Environmental Philosophy: An Introduction* (Malden, MA: Polity Press, 2015).

22   **"We place ourselves at the top":** Elizabeth Marshall Thomas, cover blurb for *Animal Wise: How We Know Animals Think and Feel,* by Virginia Morell (New York: Crown, 2013).

23   **"catching on," "making sense of things":** Linda S. Gottfredson, "Mainstream Science on Intelligence: An Editorial with 52 Signatories, History, and Bibliography," *Intelligence* 24, no. 1 (1997): 13–23.

23   **"for naming the chimpanzees":** Jane Goodall, foreword to *The Emotional Lives of Animals: A Leading Scientist Explores Animal Joy, Sorrow, and Empathy—and Why They Matter,* by Marc Bekoff (Novato, CA: New World Library, 2007), xii.

24   **"Scientists have been led to believe":** Elizabeth Marshall Thomas, *The Hidden Life of Dogs* (Boston: Houghton Mifflin, 1993), ix.

25 **"Implying that similarities exist"**: Alan M. Beck, "The Common Qualities of Man and Beast." *The Chronicle of Higher Education* (1996): n. pag. Web. Accessed 9 June 2014.

25 **"has enjoyed curious staying power"**: Alex Halberstadt. "Zoo Animals and Their Discontents." *The New York Times* 3 July 2014, Magazine sec.: n. pag. Print.

26 **"Elephant has subtle and rich feelings"**: "Baby Elephant Dives in Mud." GoWeirdFacts. http://www.goweirdfacts.com/baby-elephant-dives-in-mud.html. Accessed August 12, 2014.

26 **"because sometimes you just need to see a video of a cat petting a pig to sleep"**: Fox2 Detroit. https://www.facebook.com/WJBKFox2Detroit/videos/10153686656707994/

26 **reunion of Christian the lion:** Beckmann Group. "Christian the Lion—the full story (in HQ) [minus 20 million views!!], January 25, 2009, www.youtube.com/watch?v=md2CW4qp9e8.

CHAPTER 4: KAREN & PIPER

72 **Top Ten list:** Stanley Coren, *The Intelligence of Dogs: A Guide to the Thoughts, Emotions, and Inner Lives of Our Canine Companions* (New York: Atria, 2006), excerpt online at www.stanleycoren.com/e_intelligence.htm.

77 **"An anthropomorphic trap"**: Bruce Blumberg and Raymond Coppinger, "Can Dogs Think? Maybe Yes, and Maybe No. What Dogs Do Quite Well, Though, Is Make People Think That Dogs Can Think," *Natural History* 1 (Feb. 2005), Book Review section.

79 **"Behaviorism was an exciting adventure"**: George A. Miller, "The Cognitive Revolution: A Historical Perspective," *Trends in Cognitive Sciences* 7, no. 3 (2003): 141–44.

80 **"Exactly *how* the physical brain"**: Hanson, Rick. *Just One Thing: Developing a Buddha Brain One Simple Practice at a Time.* Oakland, CA: New Harbinger Publications, 2011. Print. Page2.

80 **"What dogs do quite well, though"**: Blumberg and Coppinger, "Can Dogs Think?"

81 **"[Charles] Darwin used observations"**: Marc Bekoff, *The Emotional Lives of Animals: A Leading Scientist Explores Animal Joy, Sorrow, and Empathy—and Why They Matter* (Novato, CA: New World Library, 2007), 32.

81   **"was comfortable with the assertion that animals have thoughts"**: Carol Kaesuk Yoon, "Donald R. Griffin, 88, Dies; Argued Animals Can Think," *The New York Times*, November 13, 2003.

84   **"three pounds of tofu-like tissue"**: Rick Hanson and Richard Mendius, *Buddha's Brain: The Practical Neuroscience of Happiness, Love & Wisdom* (Oakland, CA: New Harbinger Publications, 2009), 2.

84   **"far outside the realm"**: Yoon, "Donald R. Griffin, 88, Dies."

84   **"Wasn't it possible . . . that a chimpanzee"**: Margaret Talbot, "Birdbrain: The Woman Behind the World's Chattiest Parrots," *The New Yorker*, May 12, 2008.

84   **"Many scientists say the only reason"**: Yoon, "Donald R. Griffin, 88, Dies."

85   **"Only humans have human minds"**: Carl Safina, *Beyond Words: What Animals Think and Feel* (New York: Henry Holt, 2015).

85   **"A growing body of evidence shows"**: Frans de Waal, "The Brains of the Animal Kingdom," *The Wall Street Journal*, March 22, 2013, The Saturday Essay.

86   **"neuroanatomical, neurochemical, and neurophysiological substrates"**: Philip Low, eds. Jaak Panksepp, Diana Reiss, et al., "The Cambridge Declaration on Consciousness," publicly proclaimed in Cambridge, UK, Francis Crick Memorial Conference on Consciousness in Human and non-Human Animals, July 7, 2012.

86   **"Birds appear to offer"**: Low, "The Cambridge Declaration on Consciousness."

87   **Octopuses are . . . "social learners"**: G. Fiorito and P. Scotto, "Observational Learning in Octopus Vulgaris," Science 256 (1992): 545–46.

87   **"the Einstein of Invertebrates"**: "Octopus Tool Use: The World's Smartest Invertebrate," Science Channel, January 23, 2010, www.youtube.com/watch?v=AP_dpbTbess.

87   **"There may be no holding back its formidable intelligence"**: G. Fiorito and P. Scotto, "A demonstration of an octopus learning through observation," Discovery Channel documentary, Stazione Zoologica, Naples, Italy, www.youtube.com/watch?v=GQwJXvlTWDw;

87   **"Fish develop cultural traditions"**: Culum Brown, "Fish Intelligence, Sentience and Ethics," *Animal Cognition* 18, no. 1 (2015): 1–17.

87   **"Honeybees . . . can count":** Queen Mary University of London, "Bigger Not Necessarily Better, When It Comes to Brains," *Science Daily,* November 8, 2009, www.sciencedaily.com/releases/2009/11/091117124009.htm.

88   **"Every time someone declares that they've found *the* skill":** Virginia Morell, *Animal Wise: How We Know Animals Think and Feel* (New York: Crown Publishers, 2013), 36.

88   **"the language of the emotions":** Charles Darwin. *The Expression of the Emotions in Man and Animals* (CreateSpace Independent Publishing Platform, 2012), 151.

88   **"Human emotions, while possibly unique":** Bradshaw, John. *Dog Sense: How the New Science of Dog Behavior Can Make You a Better Friend to Your Pet.* New York: Basic Books, 2011. Print. P. 153.

89   **"Advanced neurological and genetic research":** Charles Siebert, "Should a Chimp Be Able to Sue Its Owner?" *The New York Times Magazine,* April 26, 2014.

89   **"of course dogs have feelings":** Jeffrey Moussaieff Masson. *Reflections on the Emotional World of Dogs.* New York: Crown Publishers, Inc. 1997. 3. Print.

90   **"Dolphins chuckle when they are happy":** Mark Bekoff. *The Emotional Lives of Animals: A Leading Scientist Explores Animal Joy, Sorrow, and Empathy—and Why They Matter.* Novato, CA: New World Library, 2007. 54. Print.

90   **"Even hens love to play":** Ibid., 54.

91   **"Scientists generally agree that [human] consciousness is much more complex":** Bradshaw, John. *Dog Sense: How the New Science of Dog Behavior Can Make You a Better Friend to Your Pet.* New York: Basic, 2011. Print. 212.

CHAPTER 7: KAREN & BEN

132   **"Because Florida":** Gianni Jaccoma, "This raccoon rode an alligator, because Florida." *THRILLIST/Travel.* Richard Jones, photographer. https://www.thrillist.com/travel/nation/raccoon-rides-alligator-on-florida-s-oaklawaha-river-photo Web. Accessed July 3, 2015.

133   **"I've never seen anything like it":** Associated Press, "Hamster, Snake Best Friends at Tokyo Zoo," Weird News on NBC News, January 24, 2006.

140   **"The opposite of depression":** Andrew Solomon, *The Noonday Demon: An Atlas of Depression* (New York: Scribner, 2001), 443.

CHAPTER 8: LOGAN

152   **"For a fashionable woman in Victorian England":** "Pets in the Victorian Home—Miniature Dogs," *Victoriana Magazine,* 1996, accessed June 5, 2015, www.victoriana.com/VictorianPeriod/victorian-dogs.html.

153   **"Nobody who is anybody can afford to be followed about":** Gordon Stables, "Breeding and Rearing for Pleasure, Prizes, and Profit," in *The Dog Owners' Annual for 1896* (London: Dean and Son, 1896), 166, cited in "Pride and Pedigree: The Evolution of the Victorian Dog Fancy," by Harriet Ritvo. *Victorian Studies* 29, no. 2 (Winter 1986): 227–53.

153   **"part of a new hierarchy of specialist breeders":** J. K. Walton, "Mad Dogs and Englishmen: The Conflict over Rabies in Late Victorian England," *Journal of Social History* 13, no. 2 (1979): 219–39

153   **"Somebody would show up on the promenade":** David Hancock, "What the Victorians Did for Dogs," *David Hancock on Dogs,* accessed June 7, 2015, www.davidhancockondogs.com/archives/archive_399_493/461.html.

153   **"By ten thousand years ago, dog-keeping":** John Bradshaw, *Dog Sense: How the New Science of Dog Behavior Can Make You a Better Friend to Your Pet* (New York: Basic Books, 2011), 59.

154   **"They were selecting dogs":** Rosie Cima, "Endangered Dog Breeds and the Market Forces Behind Them," *Priceonomics,* April 8, 2015.

154   **"The Victorians . . . made a dog show":** Andrzej Piotrowski, *Architecture of Thought* (Minneapolis: University of Minnesota, 2011), 194.
The Victorian-era obsession with appearance over function would end with tragic consequences for many of today's registered, purebred, show-ring dogs, crippled by exaggerated features and inherited diseases. While there are always conscientious breeders concerned with animal health, most kennel clubs and breed associations have resisted all efforts to relax their breed standards

and expand their gene pools. "In the 1950s, most breeds still had a healthy range of genetic variation," writes Dr. Bradshaw. "By 2000, only some 20 to 25 generations later, many had been inbred to the point where hundreds of genetically based deformities, diseases, and disadvantages had emerged, potentially compromising the welfare of every purebred dog." Bradshaw, *Dog Sense*. xix.

155   **"Uncle Joe and I finally decided Smoky was a crossbred Italian greyhound":** Raymond Coppinger and Lorna Coppinger. *Dogs: A Startling New Understanding of Canine Origin, Behavior, and Evolution.* (New York: Scribner, 2001), 70.

155   **"She has been depending on how one holds her in the light":** Ann Patchett. *This is the Story of A Happy Marriage.* (New York: Deckle Edge/Simon & Schuster, 2013, 77.

157   **"the rich palette of predators and scavengers":** Wolfgang M. Schleidt and Michael D. Shalter, "Co-evolution of Humans and Canids. An Alternative View of Dog Domestication: Homo Homini Lupus?" *Evolution and Cognition* 9, no. 1 (2003): 63.

157   **"a vast playground for *Canis* evolution":** Xiaoming Wang, Richard H. Tedford, and Mauricio Antón, *Dogs: Their Fossil Relatives and Evolutionary History* (New York: Columbia University Press, 2008).

157   **"an array of creatures more bizarre":** R. Dale Guthrie, "Mammals of the Mammoth Steppe as Paleoenvironmental Indicators," in *Paleoecology of Beringia*, ed. David M. Hopkins et al. (New York: Academic Press, 1982), 307.

157   **"Where the Wild Things Were":** Daniel Cossins, "Where the Wild Things Were," *The Scientist*, May 1, 2014, www.the-scientist.com/?articles.view/articleNo/39799/title/Where-the-Wild-Things-Were/.

157   **"the Wolf Event":** Wang, Tedford, and Mauricio Antón, *Dogs;* Schleidt and Shalter, "Co-evolution of Humans and Canids": 61.

157   **"All members share food and parental care":** Schleidt and Shalter, "Co-evolution of Humans and Canids": 61.

158   **the *billion* dogs:** "15,000 Years Ago, Probably in Asia, the Dog Was Born," *The New York Times*, October 19, 2015, Science section; Schleidt and Shalter, "Co-evolution of Humans and Canids": 57–72.

158   **"intentional creations of human ingenuity":** Schleidt and Shalter.

159   **adding central Asia to the atlas:** Ed Yong, "A Genetic Study Writes a New Origin Story for Dogs," *The Atlantic*, October 2015, n.p.

159   **"festooned . . . with padded armour":** N. H. Mallett, "The Dogs of War—A Short History of Canines in Combat," *Military History Now,* November 8, 2012, http://militaryhistorynow.com/2012/11/08/the-dogs-of-war-a-short-history-of-canines-in-combat/.

160   **"Dixie dingoes":** Cy Brown, "A Carolina Dog." The Bitter Southerner, 2015. http://bittersoutherner.com/carolina-dogs/#.VnLj_7YrLUI. Accessed January 12, 2015.

160   **"Many modern North American dogs":** Rhitu Chaterjee, "Barking Up the Family Tree: American Dogs Have Surprising Genetic Roots," NPR.org, last updated July 10, 2013, www.npr.org/2013/07/10/200498354/barking-up-the-family-tree-american-dogs-have-surprising-genetic-roots.

161   **"inherited a portion of their genes":** James Gorman, "Family Tree of Dogs and Wolves Is Found to Split Earlier than Thought," *The New York Times,* May 22, 2015, Science section, A10.

CHAPTER 9: HERO DOGS

167   **"driving not a Land Rover":** Alan Beck, *The Ecology of Stray Dogs: A Study of Free-Ranging Urban Animals* (West Lafayette, Indiana: Purdue University Press, NotaBell ed., 2002), Back Cover.

167   **"a sanitation inspector, dog catcher, newspaper photographer":** Ibid., 3.

167   **"In the early morning hours":** Alan Beck, "Packs of Stray Dogs Part of the Brooklyn Scene," *The New York Times,* November 12, 1972.

168   **"Man is very much part of the ecology":** Beck, *The Ecology of Stray Dogs.* 16.

168   **"the rediscovery of the dog":** Ádám Miklósi, *Dog Behaviour, Evolution, and Cognition* (Oxford: Oxford University Press, 2007), 8.

169   **"oral history of dogs":** Vilmos Csányi, *If Dogs Could Talk: Exploring the Canine Mind,* trans. Richard E. Quandt (New York: North Point, 2005). 3.

169   **the modern Iditarod race commemorates the serum-run:** Christopher Klein, "The Sled Dog Relay That Inspired the Iditarod," History.com, March 10, 2014.

170   **"While he spoke / an old hound, lying near":** Homer, *Odyssey,* trans. Robert Fitzgerald (New York: Vintage Classics, 1989), 320–21.

171   **"did not seem to be much of an improvement"**: Miklósi, *Dog Behaviour,* 8.

171   **"Pointing is one of the most widely used human nonverbal gestures"**: Ibid., 9.

173   **"spontaneously understanding another's intended"**: Brian Hare and Vanessa Woods, *The Genius of Dogs: How Dogs Are Smarter than You Think* (New York: Dutton, 2013), 37.

173   **"Everybody's dog can do calculus"**: Ibid., 38.

177   **"Do Dogs (*Canis familiaris*) Seek Help"**: Krista Macpherson and William A. Roberts, "Do Dogs (*Canis familiaris*) Seek Help in an Emergency?" *Journal of Comparative Psychology* 120, no. 2 (2006): 113–19, accessed October 8, 2015.

CHAPTER 10: IYAL

195   **"The dog was created especially"**: Henry Ward Beecher, *Proverbs from Plymouth Pulpit,* ed. William Drysdale (New York: D. Appleton and Company, 1887).

CHAPTER 11: SHELTER DOGS

199   **Close to 8 million companion animals:** "Pet Statistics," American Society for the Prevention of Cruelty to Animals, accessed October 31, 2015, www.aspca.org/about-us/faq/pet-statistics.

203   **"attention to attention"** Alexandra Horowitz, "Attention to Attention in Domestic Dog (Canis familiaris) Dyadic Play," *Animal Cognition* 12, no. 1 (2009): 107–18. Accessed August 11, 2015.

204   **In one scenario, a dog led into a room:** M. Gacsi, A. Miklósi, and O. Varqa, "Are Readers of Our Face Readers of Our Minds? Dogs (Canis Familiaris) Show Situation-dependent Recognition of Human's Attention," *Animal Cognition* 4 (2004): 144–53.

208   **Service dog work is challenging!:** Debi Davis, "Service and Therapy Work: Can One Dog Do Both?" *Alert,* National Service Dog Center newsletter, vol. 10, no. 1 (1999).

213   **An oft-told story:** Earl Hamner, Jr., and Rod Serling, "The Hunt," *The Twilight Zone,* aired January 26, 1962.

CHAPTER 12: PRISON DOGS

215 **"pretty damn worthless":** Eddie Hill, personal interview with author, July 16, 2012.

216 **In May 1989:** All the details of the crime come from the sworn testimony of witnesses who appeared in the Ohio Court of Common Pleas and testified in the case of State v. Hill, 89 CR 38 (Ct. Comm. Pl. 1989), as well as from the appellate decision, which is reported at State v. Palmer, N.E. 2d, WL 979228 (Ohio Ct. App. 1999).

217 *That gun is ridiculous:* Hill, personal interview.

217 **On Monday, May 8, 1989:** State v. Palmer.

221 **life in prison, with parole possible:** Fred Connors, "Palmer: Hill Not Guilty; Says Partner Was No Killer," *The Intelligencer/Wheeling News-Register,* August 26, 2012.

221 **"I had no idea what to expect":** Hill, personal interview.

224 **"drastically reduced incidents of violence":** Arnold Arluke and Clinton Sanders, *Between the Species: Readings in Human-Animal Relations* (Boston: Allyn & Bacon, 2008), 295.

225 **"the cats were allowed to stay":** Stacy E. Smith, "Cats and Convicts Couple Up," *PawPrints* magazine. Accessed April 29, 2013.

225 **"There was a guy killed in here":** Ibid.

226 **"The frogs are entrusted to prisoners as eggs":** Dawn Stover. "Captive Breeding". *Conservation.* The University of Washington, 5 Dec 2011. Web. 14 Oct 2015.

226 **in over two hundred prisons:** Lisa Rogak, *Dogs of Courage: The Heroism and Heart of Working Dogs Around the World* (New York: St. Martin's, 2012), 152.

226 **two-thirds involve dogs:** G. Furst, "Prison-Based Animal Programs: A National Survey," *The Prison Journal* 86, no. 4 (2006): 407–30.

226 **One and a half percent involve cats:** Ibid.

228 **71 percent of women:** "Facts About Animal Abuse & Domestic Violence," American Humane Association, n.d., accessed May 10, 2013, www.americanhumane.org/interaction/support-the-bond/fact-sheets/animal-abuse-domestic-violence.html.

228 **"Between 18 and 48 percent":** Marcotte, Amanda. "Congress Considers Bill to Help Domestic Violence Victims Escape With Their Pets." *Slate.* Slate, 15 Apr 2015. Web. 09 Mar 2016.

229   **"heart attack patients with dogs"**: Jane Weaver, "Puppy Love—It's
        Better than You Think," NBCNews.com, last updated April 8, 2004,
        www.nbcnews.com/id/4625213/ns/health-pet_health/t/puppy-love-
        ---its-better-you-think/.

229   **twenty-four stockbrokers**: Lois Baker, "Pet Dog or Cat Controls
        Blood Pressure Better than ACE Inhibitor, UB Study of Stockbrokers
        Finds," University at Buffalo, November 7, 1999, www.buffalo.edu/
        news/releases/1999/11/4489.html.

229   **"fewer allergies and less asthma"**: Weaver, "Puppy Love."

230   **In 2004, researchers at the University of Missouri**: Jane Weaver,
        "Puppy Love—It's Better than You Think," NBCNews.com, last
        updated April 8, 2004, www.nbcnews.com/id/4625213/ns/health-
        pet_health/t/puppy-love----its-better-you-think/.

230   **"Elderly people, living alone"**: Jane Goodall, foreword to *The
        Emotional Lives of Animals: A Leading Scientist Explores Animal Joy*,
        by Marc Beckoff (Novato, CA: New World Library, 2007), xiv.

230   **Scientists in Japan just announced**: Tara Haelle, "Humans Can't
        Resist Those Puppy-Dog Eyes," *HealthDay*, April 16, 2015, http://
        consumer.healthday.com/general-health-information-16/pets-and-
        health-news-531/humans-can-t-resist-those-puppy-dog-eyes-study-
        confirms-698480.html; Weaver, "Puppy Love."

230   **"love is a name"**: Carl Safina, *Beyond Words: What Animals Think
        and Feel* (New York: Henry Holt, 2015), 54.

231   **"Love helps commit us to them . . ."**: Weaver, "Puppy Love."

231   **"The warm feeling we get from our dogs"**: Alan Beck and Aaron
        Katcher, *Between Pets and People: The Importance of Animal
        Companionship* (New York: Putnam, 1983), 29.

231   **truly ancient biological connection**: Haelle, "Humans Can't Resist
        Those Puppy-Dog Eyes"; Weaver, "Puppy Love."

231   **"When people face real adversity"**: Beck and Katcher, 29.

231   **"I'm not as stupid"**: Gennifer Furst, "How Prison-Based Animal
        Programs Change Prisoner Participants," in *Between the Species,* by
        Arluke and Sanders, 293–302.

231   **"Human subpopulations"**: Ibid.

CHAPTER 13: LUCY & JOLLY

235   **Military veterans returning:** "PTSD: A Growing Epidemic," *NIH Medline Plus*, vol. 4, no. 1 (Winter 2009): 10–14. Accessed October 16, 2015.

235   **many are exploring alternative treatments:** James Dao, "Loyal Companion Helps a Veteran Regain Her Life After War Trauma," *The New York Times*, April 28, 2012.

236   **"Veterans rely on their dogs":** Janie Lorber, "For the Battle-Scarred, Comfort at Leash's End," *The New York Times*, April 3, 2010.

236   **"Skeptics say":** Dao, "Loyal Companion Helps a Veteran."

245   **"A paw on the leg":** Michaeleen Doucleff, "How Dogs Read Our Moods: Emotion Detector Found in Fido's Brain," NPR.org, last updated February 21, 2014, accessed October 16, 2015, www.npr.org/sections/health-shots/2014/02/21/280640267/how-dogs-read-our-moods-emotion-detector-found-in-fidos-brain.

246   **"Like people, dogs use simple acoustic parameters":** Attila Andics et al., "Voice-Sensitive Regions in the Dog and Human Brain Are Revealed by Comparative fMRI," *Current Biology* 2014 Mar 3;24(5):574-8. doi: 10.1016/j.cub.2014.01.058. Epub 2014 Feb 20. Accessed July 2014.

CHAPTER 14: CASEY & CONNOR II

255   **"The only purpose of the task":** University of Veterinary Medicine—Vienna. "Dogs Give Friends Food." Phys.org, December 16, 2015. http://phys.org/news/2015-12-dogs-friends-food.html. Web. Accessed December 17, 2015.

256   **"It is hard to believe that about ten years ago":** Frans de Waal. Frans de Waal-Public Page. https://www.facebook.com/Frans-de-Waal-Public-Page-99206759699/

257   **"empathic-like responding":** Deborah Custance, J.Mayer. "Empathic-like responding by domestic dogs (Canis familiaris) to distress in humans: an exploratory study." *Animal Cognition*, September 2012. 15(5):851-9. doi: 10.1007/s10071-012-0510-1. Epub 2012 May 29. http://www.ncbi.nlm.nih.gov/pubmed/22644113

258 **"In the same manner that young humans show empathy":** Stanley Coren, "Canine Empathy: Your Dog Really Does Care if You Are Unhappy," *Psychology Today,* June 7, 2012.

261 **"neurological evidence of emotions":** Gregory Berns, "Dogs Are People, Too," *The New York Times,* October 6, 2013, SR5.

262 **"Rich in dopamine receptors":** Ibid.

262 **"a natural interspecies bond":** Gregory S. Berns, Andrew M. Brooks, and Mark Spivak, "Scent of the Familiar: An fMRI Study of Canine Brain Responses to Familiar and Unfamiliar Human and Dog Odors," *Behavioural Processes* 110 (2015): 37–46.

263 **"Do these findings prove that dogs love us?":** Berns, "Dogs Are People, Too."

266 **"Almost no scientific research has been carried out":** Barbara J. King, *How Animals Grieve* (Chicago: University of Chicago, 2013). 24.

266 **"36 percent of dogs ate less than usual":** "Do Dogs Mourn?" PetPlace Staff, Easley Animal Hospital. http://www.easleyanimalhospital.com/sites/site-4631/documents/Do%20Dogs%20Mourn.pdf. Accessed May 12, 2015.

267 **"With Odin now gone":** Stanley Coren, "Do Dogs Grieve Over a Lost Loved One?" *Psychology Today,* November 18, 2014.

267 **"it's unknown how grief looks in the human brain":** Mott, "Dogs May Mourn as Deeply as Humans Do."

267 **"It's bad biology to argue against":** Bekoff, *The Emotional Lives of Animals,* xx–xxi.

268 **"To endow animals with human emotions":** Frans de Waal, "Are We in Anthropodenial?" *Discover,* July 1997, 50–53. Accessed July 30, 2011.

CHAPTER 15: LOGAN & JUKE

279 **"Dogs, dogs, and more dogs":** Cristina M. Giannantonio and Amy E. Hurley-Hanson, eds., *Extreme Leadership: Leaders, Teams and Situations Outside the Norm* (Northampton, MA, and Cheltenham, UK: Edward Elgar, 2013), 26.

CHAPTER 16:
EDDIE HILL & TIMBER & DANTE & KEEPER & JIMINY

285 **"To Timber's Trainer":** Personal letter provided by Eddie Hill.

286 **On September 20, 2012:** "Donald L. 'Duke' Palmer Jr. #1305," Clark County Prosecutor, 2012, accessed November 1, 2013, www.clarkprosecutor.org/html/death/US/palmer1305.htm.

286 **"There is . . . another victim in this case":** Fred Connors, "Palmer: Hill Not Guilty; Says Partner Was No Killer," *The Intelligencer/ Wheeling News-Register,* August 26, 2012.

Photograph Credits

PAGE 2 Samuel, Lily. *Zizou in dishwasher.* August 2, 2013. Photograph. Atlanta, Georgia.

PAGE 13 Courtesy of Donna & Jeff Erickson.

PAGE 15 Greene, Melissa Fay. *Juke.* March 4, 2013. Photograph. Unalakleet, Alaska.

PAGE 29 Caldwell, Abagail. *Schwenker twins & Barkley.* November 2013. Photograph. Marietta, Georgia.

PAGE 41 Greene, Melissa Fay. *Karen Shirk in Unalakleet.* March 8, 2013. Photograph. Unalakleet, Alaska.

PAGE 43 Gilbertson, Ashley/VII. *Trainer Jessa Brown & 4 Paws dog practice behavior disruption.* October 5, 2011. Photograph. Xenia, Ohio.

PAGE 46 Photo courtesy of Beverly Westerkamm-Wallrauch.

PAGE 51 Photo courtesy of Beverly Westerkamm-Wallrauch.

PAGE 57 *Connor, post-trach.* Courtesy of Deb & Scott Millard.

PAGE 64 *Connor in Crazy Coupe.* Photo courtesy of Deb & Scott Millard.

PAGE 69 Photo courtesy of Beverly Westerkamm-Wallrauch.

PAGE 71 Gilbertson, Ashley/VII. "Karen Shirk, executive director of 4 Paws for Ability, a service-dog school in Xenia, Ohio." *New York Times,* February 2, 2012. Photograph. Web. October 15, 2015.

PAGE 81 Samuel, Alyssa Kapnik. *Henry & Charlie in the backyard.* October 2015. Photograph. Atlanta, Georgia.

PAGE 95 Photo courtesy of Foudeelau.

PAGE 109 *Connor in Shades.* Photo courtesy of Deb and Scott Millard.

Index